Introduction to Statistics for Biomedical Engineers

Introduction to Statistics for Biomedical Engineers

Kristina M. Ropella

ISBN: 978-3-031-00492-6   paperback
ISBN: 978-3-031-01620-2   ebook

DOI 10.1007/978-3-031-01620-2

A Publication in the Springer series
SYNTHESIS LECTURES ON BIOMEDICAL ENGINEERING #14

Lecture #14

Series Editor: John D. Enderle, University of Connecticut

Series ISSN

ISSN 1930-0328 print
ISSN 1930-0336 electronic

Introduction to Statistics
for Biomedical Engineers

Kristina M. Ropella
Department of Biomedical Engineering
Marquette University

SYNTHESIS LECTURES ON BIOMEDICAL ENGINEERING #14

*This text is dedicated to all the students who have completed my BIEN 084
statistics course for biomedical engineers and have taught me how to be
more effective in communicating the subject matter and making statistics
come alive for them. I also thank J. Claypool for his patience and
for encouraging me to finally put this text together.
Finally, I thank my family for tolerating my time at home on the laptop.*

ABSTRACT

There are many books written about statistics, some brief, some detailed, some humorous, some colorful, and some quite dry. Each of these texts is designed for a specific audience. Too often, texts about statistics have been rather theoretical and intimidating for those not practicing statistical analysis on a routine basis. Thus, many engineers and scientists, who need to use statistics much more frequently than calculus or differential equations, lack sufficient knowledge of the use of statistics. The audience that is addressed in this text is the university-level biomedical engineering student who needs a bare-bones coverage of the most basic statistical analysis frequently used in biomedical engineering practice. The text introduces students to the essential vocabulary and basic concepts of probability and statistics that are required to perform the numerical summary and statistical analysis used in the biomedical field. This text is considered a starting point for important issues to consider when designing experiments, summarizing data, assuming a probability model for the data, testing hypotheses, and drawing conclusions from sampled data.

A student who has completed this text should have sufficient vocabulary to read more advanced texts on statistics and further their knowledge about additional numerical analyses that are used in the biomedical engineering field but are beyond the scope of this text. This book is designed to supplement an undergraduate-level course in applied statistics, specifically in biomedical engineering. Practicing engineers who have not had formal instruction in statistics may also use this text as a simple, brief introduction to statistics used in biomedical engineering. The emphasis is on the application of statistics, the assumptions made in applying the statistical tests, the limitations of these elementary statistical methods, and the errors often committed in using statistical analysis. A number of examples from biomedical engineering research and industry practice are provided to assist the reader in understanding concepts and application. It is beneficial for the reader to have some background in the life sciences and physiology and to be familiar with basic biomedical instrumentation used in the clinical environment.

KEYWORDS

probability model, hypothesis testing, physiology, ANOVA, normal distribution, confidence interval, power test

Contents

CHAPTER 1

Introduction

Biomedical engineers typically collect all sorts of data, from patients, animals, cell counters, micro-assays, imaging systems, pressure transducers, bedside monitors, manufacturing processes, material testing systems, and other measurement systems that support a broad spectrum of research, design, and manufacturing environments. Ultimately, the reason for collecting data is to make a decision. That decision may concern differentiating biological characteristics among different populations of people, determining whether a pharmacological treatment is effective, determining whether it is cost-effective to invest in multimillion-dollar medical imaging technology, determining whether a manufacturing process is under control, or selecting the best rehabilitative therapy for an individual patient.

The challenge in making such decisions often lies in the fact that all real-world data contains some element of uncertainty because of random processes that underlie most physical phenomenon. These random elements prevent us from predicting the exact value of any physical quantity at any moment of time. In other words, when we collect a sample or data point, we usually cannot predict the exact value of that sample or experimental outcome. For example, although the average resting heart rate of normal adults is about 70 beats per minute, we cannot predict the exact arrival time of our next heartbeat. However, we can approximate the likelihood that the arrival time of the next heartbeat will fall in a specific time interval if we have a good probability model to describe the random phenomenon contributing to the time interval between heartbeats. The timing of heart-beats is influenced by a number of physiological variables [1], including the refractory period of the individual cells that make up the heart muscle, the leakiness of the cell membranes in the sinus node (the heart's natural pacemaker), and the activity of the autonomic nervous system, which may speed up or slow down the heart rate in response to the body's need for increased blood flow, oxygen, and nutrients. The sum of these biological processes produces a pattern of heartbeats that we may measure by counting the pulse rate from our wrist or carotid artery or by searching for specific QRS waveforms in the ECG [2]. Although this sum of events makes it difficult for us to predict exactly when the new heartbeat will arrive, we can guess, with a certainty amount of confidence when the next beat will arrive. In other words, we can assign a probability to the likelihood that the next heartbeat will arrive in a specified time interval. If we were to consider all possible arrival times and assigned a probability to those arrival times, we would have a probability model for the heartbeat intervals. If we can find a probability model to describe the likelihood of occurrence of a certain event or experimental outcome, we can use statistical methods to make decisions. The probability models describe characteristics of the population or phenomenon being studied. Statistical analysis then makes use of these models to help us make decisions about the population(s) or processes.

The conclusions that one may draw from using statistical analysis are only as good as the underlying model that is used to describe the real-world phenomenon, such as the time interval between heartbeats. For example, a normally functioning heart exhibits considerable variability in beat-to-beat intervals (Figure 1.1). This variability reflects the body's continual effort to maintain homeostasis so that the body may continue to perform its most essential functions and supply the body with the oxygen and nutrients required to function normally. It has been demonstrated through biomedical research that there is a loss of heart rate variability associated with some diseases, such as diabetes and ischemic heart disease. Researchers seek to determine if this difference in variability between normal subjects and subjects with heart disease is significant (meaning, it is due to some underlying change in biology and not simply a result of chance) and whether it might be used to predict the progression of the disease [1]. One will note that the probability model changes as a consequence of changes in the underlying biological function or process. In the case of manufacturing, the probability model used to describe the output of the manufacturing process may change as

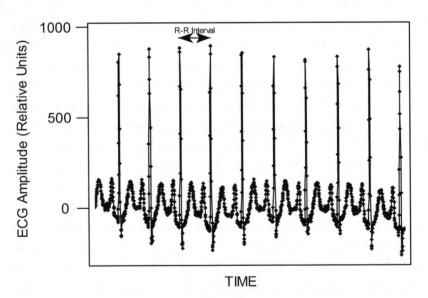

FIGURE 1.1: Example of an ECG recording, where R-R interval is defined as the time interval between successive R waves of the QRS complex, the most prominent waveform of the ECG.

a function of machine operation or changes in the surrounding manufacturing environment, such as temperature, humidity, or human operator.

Besides helping us to describe the probability model associated with real-world phenomenon, statistics help us to make decisions by giving us quantitative tools for testing hypotheses. We call this *inferential statistics*, whereby the outcome of a statistical test allows us to draw conclusions or make inferences about one or more populations from which samples are drawn. Most often, scientists and engineers are interested in comparing data from two or more different populations or from two or more different processes. Typically, the default hypothesis is that there is no difference in the distributions of two or more populations or processes, and we use statistical analysis to determine whether there are true differences in the distributions of the underlying populations to warrant different probability models be assigned to the individual processes.

In summary, biomedical engineers typically collect data or samples from various phenomena, which contain some element of randomness or unpredictable variability, for the purposes of making decisions. To make sound decisions in the context of the uncertainty with some level of confidence, we need to assume some probability model for the populations from which the samples have been collected. Once we have assumed an underlying model, we can select the appropriate statistical tests for comparing two or more populations and then use these tests to draw conclusions about

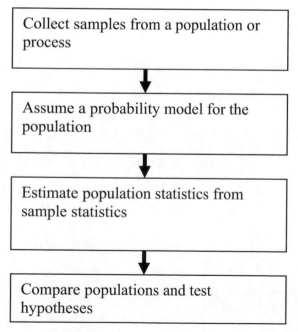

FIGURE 1.2: Steps in statistical analysis.

our hypotheses for which we collected the data in the first place. Figure 1.2 outlines the steps for performing statistical analysis of data.

In the following chapters, we will describe methods for graphically and numerically summarizing collected data. We will then talk about fitting a probability model to the collected data by briefly describing a number of well-known probability models that are used to describe biological phenomenon. Finally, once we have assumed a model for the populations from which we have collected our sample data, we will discuss the types of statistical tests that may be used to compare data from multiple populations and allow us to test hypotheses about the underlying populations.

• • • •

CHAPTER 2

Collecting Data and Experimental Design

Before we discuss any type of data summary and statistical analysis, it is important to recognize that the value of any statistical analysis is only as good as the data collected. Because we are using data or samples to draw conclusions about entire populations or processes, it is critical that the data collected (or samples collected) are representative of the larger, underlying population. In other words, if we are trying to determine whether men between the ages of 20 and 50 years respond positively to a drug that reduces cholesterol level, we need to carefully select the population of subjects for whom we administer the drug and take measurements. In other words, we have to have enough samples to represent the variability of the underlying population. There is a great deal of variety in the weight, height, genetic makeup, diet, exercise habits, and drug use in all men ages 20 to 50 years who may also have high cholesterol. If we are to test the effectiveness of a new drug in lowering cholesterol, we must collect enough data or samples to capture the variability of biological makeup and environment of the population that we are interested in treating with the new drug. Capturing this variability is often the greatest challenge that biomedical engineers face in collecting data and using statistics to draw meaningful conclusions. The experimentalist must ask questions such as the following:

- What type of person, object, or phenomenon do I sample?
- What variables that impact the measure or data can I control?
- How many samples do I require to capture the population variability to apply the appropriate statistics and draw meaningful conclusions?
- How do I avoid biasing the data with the experimental design?

Experimental design, although not the primary focus of this book, is the most critical step to support the statistical analysis that will lead to meaningful conclusions and hence sound decisions.

One of the most fundamental questions asked by biomedical researchers is, "What size sample do I need?" or "How many subjects will I need to make decisions with any level of confidence?"

We will address these important questions at the end of this book when concepts such as variability, probability models, and hypothesis testing have already been covered. For example, power tests will be described as a means for predicting the sample size required to detect significant differences in two population means using a t test.

Two elements of experimental design that are critical to prevent biasing the data or selecting samples that do not fairly represent the underlying population are randomization and blocking.

Randomization refers to the process by which we randomly select samples or experimental units from the larger underlying population such that we maximize our chance of capturing the variability in the underlying population. In other words, we do not limit our samples such that only a fraction of the characteristics or behaviors of the underlying population are captured in the samples. More importantly, we do not bias the results by artificially limiting the variability in the samples such that we alter the probability model of the sample population with respect to the probability model of the underlying population.

In addition to randomizing our selection of experimental units from which to take samples, we might also randomize our assignment of treatments to our experimental units. Or, we may randomize the order in which we take data from the experimental units. For example, if we are testing the effectiveness of two different medical imaging methods in detecting brain tumor, we will randomly assign all subjects suspect of having brain tumor to one of the two imaging methods. Thus, if we have a mix of sex, age, and type of brain tumor participating in the study, we reduce the chance of having all one sex or one age group assigned to one imaging method and a very different type of population assigned to the second imaging method. If a difference is noted in the outcome of the two imaging methods, we will not artificially introduce sex or age as a factor influencing the imaging results.

As another example, if one are testing the strength of three different materials for use in hip implants using several strength measures from a materials testing machine, one might randomize the order in which samples of the three different test materials are submitted to the machine. Machine performance can vary with time because of wear, temperature, humidity, deformation, stress, and user characteristics. If the biomedical engineer were asked to find the strongest material for an artificial hip using specific strength criteria, he or she may conduct an experiment. Let us assume that the engineer is given three boxes, with each box containing five artificial hip implants made from one of three materials: titanium, steel, and plastic. For any one box, all five implant samples are made from the same material. To test the 15 different implants for material strength, the engineer might randomize the order in which each of the 15 implants is tested in the materials testing machine so that time-dependent changes in machine performance or machine-material interactions or time-varying environmental condition do not bias the results for one or more of the materials. Thus, to fully randomize the implant testing, an engineer may literally place the numbers 1–15 in a hat and also assign the numbers 1–15 to each of the implants to be tested. The engineer will then blindly draw one of the 15 numbers from a hat and test the implant that corresponds to that number. This way the engineer is not testing all of one material in any particular order, and we avoid introducing order effects into the data.

The second aspect of experimental design is blocking. In many experiments, we are interested in one or two specific factors or variables that may impact our measure or sample. However, there may be other factors that also influence our measure and confound our statistics. In good experimental design, we try to collect samples such that different treatments within the factor of interest are not biased by the differing values of the confounding factors. In other words, we should be certain that every treatment within our factor of interest is tested within each value of the confounding factor. We refer to this design as blocking by the confounding factor. For example, we may want to study weight loss as a function of three different diet pills. One confounding factor may be a person's starting weight. Thus, in testing the effectiveness of the three pills in reducing weight, we may want to block the subjects by starting weight. Thus, we may first group the subjects by their starting weight and then test each of the diet pills within each group of starting weights.

In biomedical research, we often block by experimental unit. When this type of blocking is part of the experimental design, the experimentalist collects multiple samples of data, with each sample representing different experimental conditions, from each of the experimental units. Figure 2.1 provides a diagram of an experiment in which data are collected before and after patients receives therapy, and the experimental design uses blocking (left) or no blocking (right) by experimental unit. In the case of blocking, data are collected before and after therapy from the same set of human subjects. Thus, within an individual, the same biological factors that influence the biological response to the therapy are present before and after therapy. Each subject serves as his or her own control for factors that may randomly vary from subject to subject both before and after therapy. In essence, with blocking, we are eliminating biases in the differences between the two populations

Block (Repeated Measures)				No Block (No repeated measures)			
Subject	Measure before treatment	Measure after treatment		Subject	Measure before treatment	Subject	Measure after treatment
1	M11	12		1	M1	K+1	M(K+1)
2	M21	M22		2	M2	K+2	M(K+2)
3	M31	M32		3	M3	K+3	M(K+3)
.				.		.	
.				.		.	
K	MK1	MK2		K	MK	K+K	M(K+K)

FIGURE 2.1: Samples are drawn from two populations (before and after treatment), and the experimental design uses block (left) or no block (right). In this case, the block is the experimental unit (subject) from which the measures are made.

(before and after) that may result because we are using two different sets of experimental units. For example, if we used one set of subjects before therapy and then an entirely different set of subjects after therapy (Figure 2.1, right), there is a chance that the two sets of subjects may vary enough in sex, age, weight, race, or genetic makeup, which would lead to a difference in response to the therapy that has little to do with the underlying therapy. In other words, there may be confounding factors that contribute to the difference in the experimental outcome before and after therapy that are not only a factor of the therapy but really an artifact of differences in the distributions of the two different groups of subjects from which the two samples sets were chosen. Blocking will help to eliminate the effect of intersubject variability.

However, blocking is not always possible, given the nature of some biomedical research studies. For example, if one wanted to study the effectiveness of two different chemotherapy drugs in reducing tumor size, it is impractical to test both drugs on the same tumor mass. Thus, the two drugs are tested on different groups of individuals. The same type of design would be necessary for testing the effectiveness of weight-loss regimens.

Thus, some important concepts and definitions to keep in mind when designing experiments include the following:

- *experimental unit*: the item, object, or subject to which we apply the treatment and from which we take sample measurements;
- *randomization*: allocate the treatments randomly to the experimental units;
- *blocking*: assigning all treatments within a factor to every level of the blocking factor. Often, the blocking factor is the experimental unit. Note that in using blocking, we still randomize the order in which treatments are applied to each experimental unit to avoid ordering bias.

Finally, the experimentalist must always think about how representative the sample population is with respect to the greater underlying population. Because it is virtually impossible to test every member of a population or every product rolling down an assembly line, especially when destructive testing methods are used, the biomedical engineer must often collect data from a much smaller sample drawn from the larger population. It is important, if the statistics are going to lead to useful conclusions, that the sample population captures the variability of the underlying population. What is even more challenging is that we often do not have a good grasp of the variability of the underlying population, and because of expense and respect for life, we are typically limited in the number of samples we may collect in biomedical research and manufacturing. These limitations are not easy to address and require that the engineer always consider how fair the sample and data analysis is and how well it represents the underlying population(s) from which the samples are drawn.

· · · ·

CHAPTER 3

Data Summary and Descriptive Statistics

We assume now that we have collected our data through the use of good experimental design. We now have a collection of numbers, observations, or descriptions to describe our data, and we would like to summarize the data to make decisions, test a hypothesis, or draw a conclusion.

3.1 WHY DO WE COLLECT DATA?

The world is full of uncertainty, in the sense that there are random or unpredictable factors that influence every experimental measure we make. The unpredictable aspects of the experimental outcomes also arise from the variability in biological systems (due to genetic and environmental factors) and manufacturing processes, human error in making measurements, and other underlying processes that influence the measures being made.

Despite the uncertainty regarding the exact outcome of an experiment or occurrence of a future event, we collect data to try to better understand the processes or populations that influence an experimental outcome so that we can make some predictions. Data provide information to reduce uncertainty and allow for decision making. When properly collected and analyzed, data help us solve problems. It cannot be stressed enough that the data must be properly collected and analyzed if the data analysis and subsequent conclusions are to have any value.

3.2 WHY DO WE NEED STATISTICS?

We have three major reasons for using statistical data summary and analysis:

1. The real world is full of random events that cannot be described by exact mathematical expressions.
2. Variability is a natural and normal characteristic of the natural world.
3. We like to make decisions with some confidence. This means that we need to find trends within the variability.

3.3   WHAT QUESTIONS DO WE HOPE TO ADDRESS WITH OUR STATISTICAL ANALYSIS?

There are several basic questions we hope to address when using numerical and graphical summary of data:

1.   Can we differentiate between groups or populations?
2.   Are there correlations between variables or populations?
3.   Are processes under control?

Finding physiological differences between populations is probably the most frequent aim of biomedical research. For example, researchers may want to know if there is a difference in life expectancy between overweight and underweight people. Or, a pharmaceutical company may want to determine if one type of antibiotic is more effective in combating bacteria than another. Or, a physician wonders if diastolic blood pressure is reduced in a group of hypertensive subjects after the consumption of a pressure-reducing drug. Most often, biomedical researchers are comparing populations of people or animals that have been exposed to two or more different treatments or diagnostic tests, and they want to know if there is difference between the responses of the populations that have received different treatments or tests. Sometimes, we are drawing multiple samples from the same group of subjects or experimental units. A common example is when the physiological data are taken before and after some treatment, such as drug intake or electronic therapy, from one group of patients. We call this type of data collection *blocking* in the experimental design. This concept of blocking is discussed more fully in Chapter 2.

Another question that is frequently the target of biomedical research is whether there is a correlation between two physiological variables. For example, is there a correlation between body build and mortality? Or, is there a correlation between fat intake and the occurrence of cancerous tumors. Or, is there a correlation between the size of the ventricular muscle of the heart and the frequency of abnormal heart rhythms? These type of questions involve collecting two set of data and performing a correlation analysis to determine how well one set of data may be predicted from another. When we speak of correlation analysis, we are referring to the linear relation between two variables and the ability to predict one set of data by modeling the data as a linear function of the second set of data. Because correlation analysis only quantifies the linear relation between two processes or data sets, nonlinear relations between the two processes may not be evident. A more detailed description of correlation analysis may be found in Chapter 7.

Finally, a biomedical engineer, particularly the engineer involved in manufacturing, may be interested in knowing whether a manufacturing process is under control. Such a question may arise if there are tight controls on the manufacturing specifications for a medical device. For example, if the engineer is trying to ensure quality in producing intravascular catheters that must have diameters between 1 and 2 cm, the engineer may randomly collect samples of catheters from the assembly line at random intervals during the day, measure their diameters, determine how many of the catheters meet specifications, and determine whether there is a sudden change in the number of catheters that fail to meet specifications. If there is such a change, the engineers may look for elements of the manufacturing process that change over time, changes in environmental factors, or user errors. The engineer can use control charts to assess whether the processes are under control. These methods of statistical analysis are not covered in this text, but may be found in a number of references, including [3].

3.4   HOW DO WE GRAPHICALLY SUMMARIZE DATA?

We can summarize data in graphical or numerical form. The numerical form is what we refer to as statistics. Before blindly applying the statistical analysis, it is always good to look at the raw data, usually in a graphical form, and then use graphical methods to summarize the data in an easy to interpret format.

The types of graphical displays that are most frequently used by biomedical engineers include the following: scatterplots, time series, box-and-whisker plots, and histograms.

Details for creating these graphical summaries are described in [3–6], but we will briefly describe them here.

3.4.1   Scatterplots

The scatterplot simply graphs the occurrence of one variable with respect to another. In most cases, one of the variables may be considered the independent variable (such as time or subject number), and the second variable is considered the dependent variable. Figure 3.1 illustrates an example of a scatterplot for two sets of data. In general, we are interested in whether there is a predictable relationship that maps our independent variable (such as respiratory rate) into our dependent variable (such a heart rate). If there is a linear relationship between the two variables, the data points should fall close to a straight line.

3.4.2   Time Series

A time series is used to plot the changes in a variable as a function of time. The variable is usually a physiological measure, such as electrical activation in the brain or hormone concentration in the blood stream, that changes with time. Figure 3.2 illustrates an example of a time series plot. In this figure, we are looking at a simple sinusoid function as it changes with time.

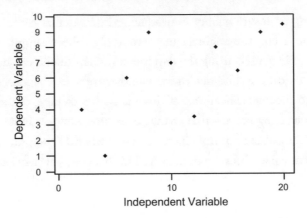

FIGURE 3.1: Example of a scatterplot.

3.4.3 Box-and-Whisker Plots

These plots illustrate the first, second, and third quartiles as well as the minimum and maximum values of the data collected. The second quartile (Q2) is also known as the median of the data. This quantity, as defined later in this text, is the middle data point or sample value when the samples are listed in descending order. The first quartile (Q1) can be thought of as the median value of the samples that fall below the second quartile. Similarly, the third quartile (Q3) can be thought of as the median value of the samples that fall above the second quartile. Box-and-whisker plots are useful in that they highlight whether there is skew to the data or any unusual outliers in the samples (Figure 3.3).

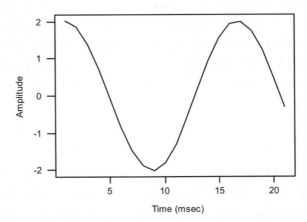

FIGURE 3.2: Example of a time series plot. The amplitude of the samples is plotted as a function of time.

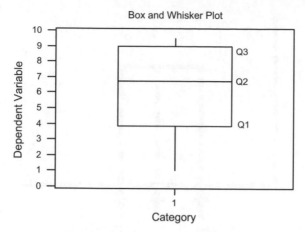

FIGURE 3.3: Illustration of a box-and-whisker plot for the data set listed. The first (Q1), second (Q2), and third (Q3) quartiles are shown. In addition, the whiskers extend to the minimum and maximum values of the sample set.

3.4.4   Histogram

The histogram is defined as a frequency distribution. Given N samples or measurements, x_i, which range from X_{min} to X_{max}, the samples are grouped into nonoverlapping intervals (bins), usually of equal width (Figure 3.4). Typically, the number of bins is on the order of 7–14, depending on the nature of the data. In addition, we typically expect to have at least three samples per bin [7]. Sturgess' rule [6] may also be used to estimate the number of bins and is given by

$$k = 1 + 3.3 \log(n).$$

where k is the number of bins and n is the number of samples.

Each bin of the histogram has a lower boundary, upper boundary, and midpoint. The histogram is constructed by plotting the number of samples in each bin. Figure 3.5 illustrates a histogram for 1000 samples drawn from a normal distribution with mean (μ) = 0 and standard deviation (σ) = 1.0. On the horizontal axis, we have the sample value, and on the vertical axis, we have the number of occurrences of samples that fall within a bin.

Two measures that we find useful in describing a histogram are the absolute frequency and relative frequency in one or more bins. These quantities are defined as

a) f_i = absolute frequency in ith bin;
b) f_i/n = relative frequency in ith bin, where n is the total number of samples being summarized in the histogram.

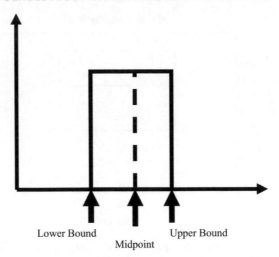

FIGURE 3.4: One bin of a histogram plot. The bin is defined by a lower bound, a midpoint, and an upper bound.

A number of algorithms used by biomedical instruments for diagnosing or detecting abnormalities in biological function make use of the histogram of collected data and the associated relative frequencies of selected bins [8]. Often times, normal and abnormal physiological functions (breath sounds, heart rate variability, frequency content of electrophysiological signals) may be differentiated by comparing the relative frequencies in targeted bins of the histograms of data representing these biological processes.

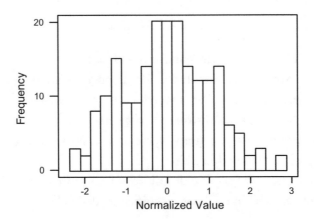

FIGURE 3.5: Example of a histogram plot. The value of the measure or sample is plotted on the horizontal axis, whereas the frequency of occurrence of that measure or sample is plotted along the vertical axis.

The histogram can exhibit several shapes. The shapes, illustrated in Figure 3.6, are referred to as symmetric, skewed, or bimodal.

A skewed histogram may be attributed to the following [9]:

1. mechanisms of interest that generate the data (e.g., the physiological mechanisms that determine the beat-to-beat intervals in the heart);
2. an artifact of the measurement process or a shift in the underlying mechanism over time (e.g., there may be time-varying changes in a manufacturing process that lead to a change in the statistics of the manufacturing process over time);
3. a mixing of populations from which samples are drawn (this is typically the source of a bimodal histogram).

The histogram is important because it serves as a rough estimate of the true probability density function or probability distribution of the underlying random process from which the samples are being collected.

The probability density function or probability distribution is a function that quantifies the probability of a random event, x, occurring. When the underlying random event is discrete in nature, we refer to the probability density function as the probability mass function [10]. In either case, the function describes the probabilistic nature of the underlying random variable or event and allows us to predict the probability of observing a specific outcome, x (represented by the random variable), of an experiment. The cumulative distribution function is simply the sum of the probabilities for a group of outcomes, where the outcome is less than or equal to some value, x.

Let us consider a random variable for which the probability density function is well defined (for most real-world phenomenon, such a probability model is not known.) The random variable is the outcome of a single toss of a dice. Given a single fair dice with six sides, the probability of rolling a six on the throw of a dice is 1 of 6. In fact, the probability of throwing a one is also 1 of 6. If we consider all possible outcomes of the toss of a dice and plot the probability of observing any one of those six outcomes in a single toss, we would have a plot such as that shown in Figure 3.7.

This plot shows the probability density or probability mass function for the toss of a dice. This type of probability model is known as a uniform distribution because each outcome has the exact same probability of occurring (1/6 in this case).

For the toss of a dice, we know the true probability distribution. However, for most real-world random processes, especially biological processes, we do not know what the true probability density or mass function looks like. As a consequence, we have to use the histogram, created from a small sample, to try to estimate the best probability distribution or probability model to describe the real-world phenomenon. If we return to the example of the toss of a dice, we can actually toss the dice a number of times and see how close the histogram, obtained from experimental data, matches

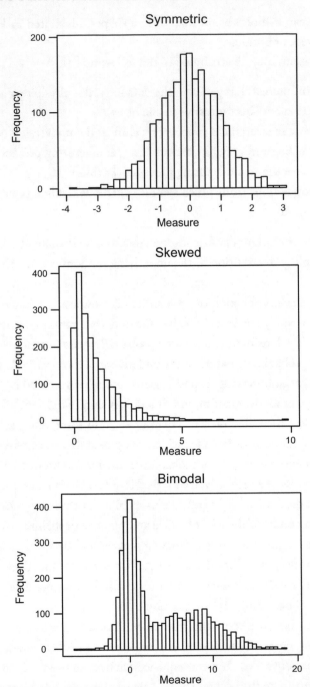

FIGURE 3.6: Examples of a symmetric (top), skewed (middle), and bimodal (bottom) histogram. In each case, 2000 sampled were drawn from the underlying populations.

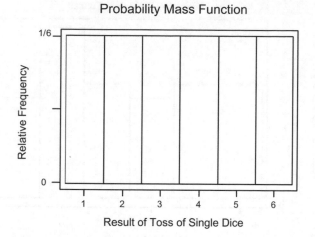

FIGURE 3.7: The probability density function for a discrete random variable (probability mass function). In this case, the random variable is the value of a toss of a single dice. Note that each of the six possible outcomes has a probability of occurrence of 1 of 6. This probability density function is also known as a uniform probability distribution.

the true probability mass function for the ideal six-sided dice. Figure 3.8 illustrates the histograms for the outcomes of 50 and 1000 tosses of a single dice. Note that even with 50 tosses or samples, it is difficult to determine what the true probability distribution might look like. However, as we approach 1000 samples, the histogram is approaching the true probability mass function (the uniform distribution) for the toss of a dice. But, there is still some variability from bin to bin that does not look as uniform as the ideal probability distribution illustrated in Figure 3.7. The message to take away from this illustration is that most biomedical research reports the outcomes of a small number of samples. It is clear from the dice example that the statistics of the underlying random process are very difficult to discern from a small sample, yet most biomedical research relies on data from small samples.

3.5   GENERAL APPROACH TO STATISTICAL ANALYSIS

We have now collected our data and looked at some graphical summaries of the data. Now we will use numerical summary, also known as statistics, to try to describe the nature of the underlying population or process from which we have taken our samples. From these descriptive statistics, we assume a probability model or probability distribution for the underlying population or process and then select the appropriate statistical tests to test hypotheses or make decisions. It is important to note that the conclusions one may draw from a statistical test depends on how well the assumed probability model fits the underlying population or process.

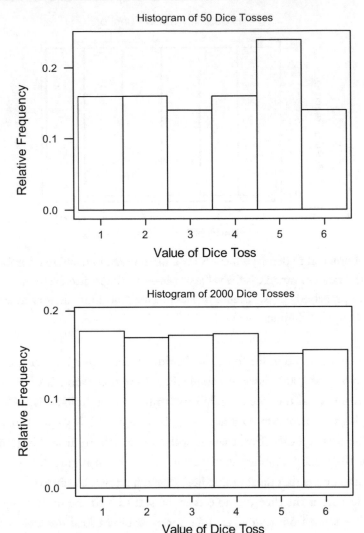

FIGURE 3.8: Histograms representing the outcomes of experiments in which a single dice is tossed 50 (top) and 2000 times (lower), respectively. Note that as the sample size increases, the histogram approaches the true probability distribution illustrated in Figure 3.7.

As stated in the Introduction, biomedical engineers are trying to make decisions about populations or processes to which they have limited access. Thus, they design experiments and collect samples that they think will fairly represent the underlying population or process. Regardless of what type of statistical analysis will result from the investigation or study, all statistical analysis should follow the same general approach:

1. Measure a limited number of representative samples from a larger population.
2. Estimate the true statistics of larger population from the sample statistics.

Some important concepts need to be addressed here. The first concept is somewhat obvious. It is often impossible or impractical to take measurements or observations from an entire population. Thus, the biomedical engineer will typically select a smaller, more practical sample that represents the underlying population and the extent of variability in the larger population. For example, we cannot possibly measure the resting body temperature of every person on earth to get an estimate of normal body temperature and normal range. We are interested in knowing what the normal body temperature is, on average, of a healthy human being and the normal range of resting temperatures as well as the likelihood or probability of measuring a specific body temperature under healthy, resting conditions. In trying to determine the characteristics or underlying probability model for body temperature for healthy, resting individuals, the researcher will select, at random, a sample of healthy, resting individuals and measure their individual resting body temperatures with a thermometer. The researchers will have to consider the composition and size of the sample population to adequately represent the variability in the overall population. The researcher will have to define what characterizes a normal, healthy individual, such as age, size, race, sex, and other traits. If a researcher were to collect body temperature data from such a sample of 3000 individuals, he or she may plot a histogram of temperatures measured from the 3000 subjects and end up with the following histogram (Figure 3.9). The researcher may also calculate some basic descriptive statistics for the 3000 samples, such as sample average (mean), median, and standard deviation.

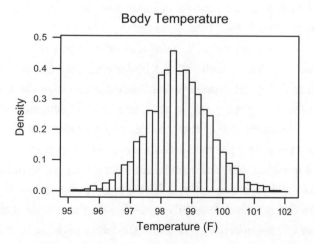

FIGURE 3.9: Histogram for 2000 internal body temperatures collected from a normally distributed population.

Once the researcher has estimated the sample statistics from the sample population, he or she will try to draw conclusions about the larger (true) population. The most important question to ask when reviewing the statistics and conclusions drawn from the sample population is how well the sample population represents the larger, underlying population.

Once the data have been collected, we use some basic descriptive statistics to summarize the data. These basic descriptive statistics include the following general measures: central tendency, variability, and correlation.

3.6   DESCRIPTIVE STATISTICS

There are a number of descriptive statistics that help us to picture the distribution of the underlying population. In other words, our ultimate goal is to assume an underlying probability model for the population and then select the statistical analyses that are appropriate for that probability model.

When we try to draw conclusions about the larger underlying population or process from our smaller sample of data, we assume that the underlying model for any sample, "event," or measure (the outcome of the experiment) is as follows:

$$X = \mu \pm \text{individual differences} \pm \text{situational factors} \pm \text{unknown variables},$$

where X is our measure or sample value and is influenced by μ, which is the true population mean; individual differences such as genetics, training, motivation, and physical condition; situation factors, such as environmental factors; and unknown variables such as unidentified/nonquantified factors that behave in an unpredictable fashion from moment to moment.

In other words, when we make a measurement or observation, the measured value represents or is influenced by not only the statistics of the underlying population, such as the population mean, but factors such as biological variability from individual to individual, environmental factors (time, temperature, humidity, lighting, drugs, etc.), and random factors that cannot be predicted exactly from moment to moment. All of these factors will give rise to a histogram for the sample data, which may or may not reflect the true probability density function of the underlying population. If we have done a good job with our experimental design and collected a sufficient number of samples, the histogram and descriptive statistics for the sample population should closely reflect the true probability density function and descriptive statistics for the true or underlying population. If this is the case, then we can make conclusions about the larger population from the smaller sample population. If the sample population does not reflect variability of the true population, then the conclusions we draw from statistical analysis of the sample data may be of little value.

There are a number of probability models that are useful for describing biological and manufacturing processes. These include the normal, Poisson, exponential, and gamma distributions [10]. In this book, we will focus on populations that follow a normal distribution because this is the most frequently encountered probability distribution used in describing populations. Moreover, the most frequently used methods of statistical analysis assume that the data are well modeled by a normal ("bell-curve") distribution. It is important to note that many biological processes are not well modeled by a normal distribution (such as heart rate variability), and the statistics associated with the normal distribution are not appropriate for such processes. In such cases, nonparametric statistics, which do not assume a specific type of distribution for the data, may serve the researcher better in understanding processes and making decisions. However, using the normal distribution and its associated statistics are often adequate given the central limit theorem, which simply states that the sum of random processes with arbitrary distributions will result in a random variable with a normal distribution. One can assume that most biological phenomena result from a sum of random processes.

3.6.1   Measures of Central Tendency

There are several measures that reflect the central tendency or concentration of a sample population: sample mean (arithmetic average), sample median, and sample mode.

The sample mean may be estimated from a group of samples, x_i, where i is sample number, using the formula below.

Given n data points, $x_1, x_2, ..., x_n$:

$$\bar{x} = \frac{1}{n} \sum_{i=1}^{n} x_i.$$

In practice, we typically do not know the true mean, μ, of the underlying population, instead we try to estimate true mean, μ, of the larger population. As the sample size becomes large, the sample mean, $\bar{x}$, should approach the true mean, μ, assuming that the statistics of the underlying population or process do not change over time or space.

One of the problems with using the sample mean to represent the central tendency of a population is that the sample mean is susceptible to outliers. This can be problematic and often deceiving when reporting the average of a population that is heavily skewed. For example, when reporting income for a group of new college graduates for which one is an NBA player who has just signed a multimillion-dollar contract, the estimated mean income will be much greater than what most graduates earns. The same misrepresentation is often evident when reporting mean value for homes in a specific geographic region where a few homes valued on the order of a million can hide the fact that several hundred other homes are valued at less than $200,000.

Another useful measure for summarizing the central tendency of a population is the sample median. The median value of a group of observations or samples, x_j, is the middle observation when samples, x_j, are listed in descending order.

For example, if we have the following values for tidal volume of the lung:

$$2, 1.5, 1.3, 1.8, 2.2, 2.5, 1.4, 1.3,$$

we can find the median value by first ordering the data in descending order:

$$2.5, 2.2, 2.0, 1.8, 1.5, 1.4, 1.3, 1.3,$$

and then we cross of values on each end until we reach a middle value:

$$\cancel{2.5}, \cancel{2.2}, \cancel{2.0}, 1.8, 1.5, \cancel{1.4}, \cancel{1.3}, \cancel{1.3}.$$

In this case, there are two middle values; thus, the median is the average of those two values, which is 1.65.

Note that if the number of samples, n, is odd, the median will be the middle observation. If the sample size, n, is even, then the median equals the average of two middle observations. Compared with the sample mean, the sample median is less susceptible to outliers. It ignores the skew in a group of samples or in the probability density function of the underlying population. In general, to fairly represent the central tendency of a collection of samples or the underlying population, we use the following rule of thumb:

1. If the sample histogram or probability density function of the underlying population is symmetric, use mean as a central measure. For such populations, the mean and median are about equal, and the mean estimate makes use of all the data.
2. If the sample histogram or probability density function of the underlying population is skewed, median is a more fair measure of center of distribution.

Another measure of central tendency is mode, which is simply the most frequent observation in a collection of samples. In the tidal volume example given above, 1.3 is the most frequently occurring sample value. Mode is not used as frequently as mean or median in representing central tendency.

3.6.2   Measures of Variability

Measures of central tendency alone are insufficient for representing the statistics of a population or process. In fact, it is usually the variability in the population that makes things interesting and leads to uncertainty in decision making. The variability from subject to subject, especially in physiological function, is what makes finding fool-proof diagnosis and treatment often so difficult. What works for one person often fails for another, and, it is not the mean or median that picks up on those subject-to-subject differences, but rather the variability, which is reflected in differences in the probability models underlying those different populations.

When summarizing the variability of a population or process, we typically ask, "How far from the center (sample mean) do the samples (data) lie?" To answer this question, we typically use the following estimates that represent the spread of the sample data: interquartile ranges, sample variance, and sample standard deviation.

The interquartile range is the difference between the first and third quartiles of the sample data. For sampled data, the median is also known as the second quartile, Q2. Given Q2, we can find the first quartile, Q1, by simply taking the median value of those samples that lie below the second quartile. We can find the third quartile, Q3, by taking the median value of those samples that lie above the second quartile. As an illustration, we have the following samples:

$$1, 3, 3, 2, 5, 1, 1, 4, 3, 2.$$

If we list these samples in descending order,

$$5, 4, 3, 3, 3, 2, 2, 1, 1, 1,$$

the median value and second quartile for these samples is 2.5. The first quartile, Q1, can be found by taking the median of the following samples,

$$2.5, 2, 2, 1, 1, 1,$$

which is 1.5. In addition, the third quartile, Q3, may be found by taking the median value of the following samples:

$$5, 4, 3, 3, 3, 2.5,$$

which is 3. Thus, the interquartile range, Q3 − Q1 = 3 − 1.5 = 2.

Sample variance, s^2, is defined as the "average distance of data from the mean" and the formula for estimating s^2 from a collection of samples, x_i, is

$$s^2 = \frac{1}{n-1} \sum_{i-1}^{n} (x_i - \bar{x})^2.$$

Sample standard deviation, s, which is more commonly referred to in describing the variability of the data is

$$s = \sqrt{s^2} \text{ (same units as original samples).}$$

It is important to note that for normal distributions (symmetrical histograms), sample mean and sample deviation are the only parameters needed to describe the statistics of the underlying phenomenon. Thus, if one were to compare two or more normally distributed populations, one only need to test the equivalence of the means and variances of those populations.

•   •   •   •

CHAPTER 4

Assuming a Probability Model From the Sample Data

Now that we have collected the data, graphed the histogram, estimated measures of central tendency and variability, such as mean, median, and standard deviation, we are ready to assume a probability model for the underlying population or process from which we have obtained samples. At this point, we will make a rough assumption using simple measures of mean, median, standard deviation and the histogram. But it is important to note that there are more rigorous tests, such as the χ^2 test for normality [7] to determine whether a particular probability model is appropriate to assume from a collection of sample data.

Once we have assumed an appropriate probability model, we may select the appropriate statistical tests that will allow us to test hypotheses and draw conclusions with some level of confidence. The probability model will dictate what level of confidence we have when accepting or rejecting a hypothesis.

There are two fundamental questions that we are trying to address when assuming a probability model for our underlying population:

1. How confident are we that the sample statistics are representative of the entire population?
2. Are the differences in the statistics between two populations significant, resulting from factors other than chance alone?

To declare any level of "confidence" in making statistical inference, we need a mathematical model that describes the probability that any data value might occur. These models are called probability distributions.

There are a number of probability models that are frequently assumed to describe biological processes. For example, when describing heart rate variability, the probability of observing a specific time interval between consecutive heartbeats might be described by an exponential distribution [1, 8]. Figure 3.6 in Chapter 3 illustrates a histogram for samples drawn from an exponential distribution.

Note that this distribution is highly skewed to the right. For R-R intervals, such a probability function makes sense physiologically because the individual heart cells have a refractory period that prevents them from contracting in less that a minimum time interval. Yet, a very prolonged time interval may occur between beats, giving rise to some long time intervals that occur infrequently.

The most frequently assumed probability model for most scientific and engineering applications is the normal or Gaussian distribution. This distribution is illustrated by the solid black line in Figure 4.1 and often referred to as the bell curve because it looks like a musical bell.

The equation that gives the probability, $f(x)$, of observing a specific value of x from the underlying *normal* population is

$$f(x) = \frac{1}{\sigma\sqrt{2\pi}} e^{-\frac{1}{2}\left(\frac{x-\mu}{\sigma}\right)^2}, \quad \infty < x < \infty$$

where μ is the true mean of the underlying population or process and σ is the standard deviation of the same population or process. A graph of this equation is given illustrated by the solid, smooth curve in Figure 4.1. The area under the curve equals one.

Note that the normal distribution is

1. a symmetric, bell-shaped curve completely described by its mean, μ, and standard deviation, σ.
2. by changing μ and σ, we stretch and slide the distribution.

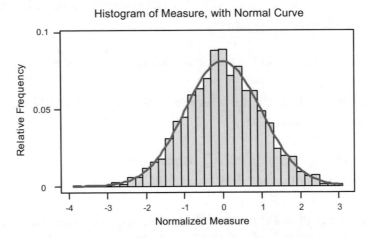

FIGURE 4.1: A histogram of 1000 samples drawn from a normal distribution is illustrated. Superimposed on the histogram is the ideal normal curve representing the normal probability distribution function.

Figure 4.1 also illustrates a histogram that is obtained when we randomly select 1000 samples from a population that is normally distributed and has a mean of 0 and a variance of 1. It is important to recognize that as we increase the sample size n, the histogram approaches the ideal normal distribution shown with the solid, smooth line. But, at small sample sizes, the histogram may look very different from the normal curve. Thus, from small sample sizes, it may be difficult to determine if the assumed model is appropriate for the underlying population or process, and any statistical tests that we perform may not allow us to test hypotheses and draw conclusions with any real level of confidence.

We can perform linear operations on our normally distributed random variable, x, to produce another normally distributed random variable, y. These operations include multiplication of x by a constant and addition of a constant (offset) to x. Figure 4.2 illustrates histograms for samples drawn from each of populations x and y. We note that the distribution for y is shifted (the mean is now equal to 5) and the variance has increased with respect to x.

One test that we may use to determine how well a normal probability model fits our data is to count how many samples fall within ±1 and ±2 standard deviations of the mean. If the data and underlying population or process is well modeled by a normal distribution, 68% of the samples should lie within ±1 standard deviation from the mean and 95% of the samples should lie within

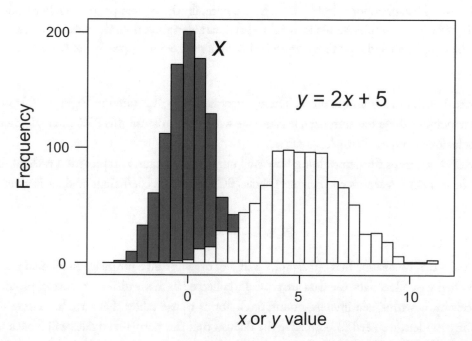

FIGURE 4.2: Histograms are shown for samples drawn from populations x and y, where y is simply a linear function of x. Note that the mean and variance of y differ from x, yet both are normal distributions.

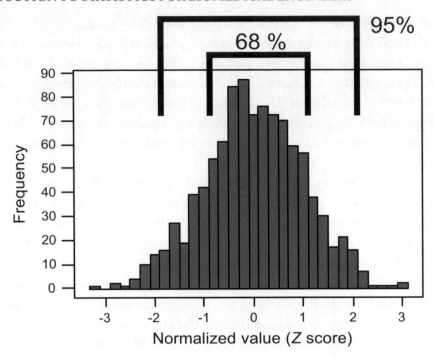

FIGURE 4.3: Histogram for samples drawn from a normally distributed population. For a normal distribution, 68% of the samples should lie within ±1 standard deviation from the mean (0 in this case) and 95% of the samples should lie within ±2 standard deviations (1.96 to be precise) of the mean.

±2 standard deviations from the mean. These percentages are illustrated in Figure 4.3. It is important to remember these few numbers, because we will frequently use this 95% interval when drawing conclusions from our statistical analysis.

Another means for determining how well our sampled data, x, represent a normal distribution is the estimate Pearson's coefficient of skew (PCS) [5]. The coefficient of skew is given by

$$\text{PCS} = \frac{3\left|\bar{x} - x_{\text{median}}\right|}{s}.$$

If the PCS > 0.5, we assume that our samples were not drawn from a normally distributed population.

When we collect data, the data are typically collected in many different types of physical units (volts, celsius, newtons, centimeters, grams, etc.). For us to use tables that have been developed for probability models, we need to normalize the data so that the normalized data will have a mean of 0 and a standard deviation of 1. Such a normal distribution is called a standard normal distribution and is illustrated in Figure 4.1.

The standard normal distribution has a bell-shaped, symmetric distribution with $\mu = 0$ and $\sigma = 1$.

To convert normally distributed data to the standard normal value, we use the following formulas,

$$z = (x - \mu)/\sigma \quad \text{or} \quad z = (x - \bar{x})/s,$$

depending on if we know the true mean, μ, and standard deviation, α, or we only have the sample estimates, $\bar{x}$ or s.

For any individual sample or data point, x_i, from a sample with mean, $\bar{x}$, and standard deviation, s, we can determine its z score from the following formula:

$$z_i = \frac{x_i - \bar{x}}{s}.$$

For an individual sample, the z score is a "normalized" or "standardized" value. We can use this value with our equations for probability density function or our standardized probability tables [3] to determine the probability of observing such a sample value from the underlying population.

The z score can also be thought of as a measure of the distance of the individual sample, x_i, from the sample average, $\bar{x}$, in units of standard deviation. For example, if a sample point, x_i has a z score of $z_i = 2$, it means that the data point, x_i, is 2 standard deviations from the sample mean.

We use normalized z scores instead of the original data when performing statistical analysis because the tables for the normalized data are already worked out and available in most statistics texts or statistical software packages. In addition, by using normalized values, we need not worry about the absolute amplitude of the data or the units used to measure the data.

4.1   THE STANDARD NORMAL DISTRIBUTION

The standard normal distribution is illustrated in Table 4.1.

The z table associated with this figure provides table entries that give the probability that $z \leq a$, which equals the area under the normal curve to the left of $z = a$. If our data come from a normal distribution, the table tells us the probability or "chance" of our sample value or experimental outcomes having a value less than or equal to a.

Thus, we can take any sample and compute its z score as described above and then use the z table to find the probability of observing a z value that is less than or equal to some normalized value, a. For example, the probability of observing a z value that is less than or equal to 1.96 is 97.5%. Thus, the probability of observing a z value greater than 1.96 is 2.5%. In addition, because of symmetry in the distribution, we know that the probability of observing a z value greater than –1.96 is also 97.5%, and the probability of observing a z value less than or equal to –1.96 is 2.5%. Finally,

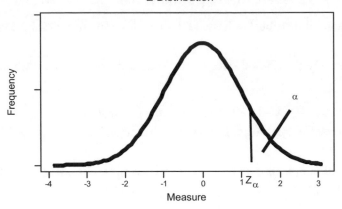

Z Distribution

Area to left of z_α equals the $\Pr(z < z_\alpha) = 1 - \alpha$; thus, the area in the tail to the right of z_α equals α.

TABLE 4.1: Standard z distribution function: areas under standardized normal density function

z	0.00	0.01	0.02	0.03	0.04	0.05	0.06	0.07	0.08	0.09
0.0	0.5000	0.5040	0.5080	0.5120	0.5160	0.5199	0.5239	0.15279	0.5319	0.5359
0.1	0.5398	0.5438	0.5478	0.5517	0.5557	0.5596	0.5636	0.5675	0.5714	0.5753
0.2	0.5793	0.5832	0.5871	0.5910	0.5948	0.5987	0.6026	0.6064	0.6103	0.6141
...										
1.7	0.9554	0.9564	0.9573	0.9582	0.9591	0.9599	0.9608	0.9616	0.9625	0.9633
1.8	0.9641	0.9649	0.9656	0.9664	0.9671	0.9678	0.9686	0.9693	0.9699	0.9706
1.9	0.9713	0.9719	0.9726	0.9732	0.9738	0.9744	0.9750	0.9756	0.9761	0.9767
2.0	0.9772	0.9778	0.9783	0.9788	0.9793	0.9798	0.9803	0.9808	0.9812	0.9817
...										
2.4	0.9918	0.9920	0.9922	0.9925	0.9927	0.9929	0.9931	0.9932	0.9934	0.9936
2.5	0.9938	0.9940	0.9941	0.9943	0.9945	0.9946	0.9948	0.9949	0.9951	0.9952
2.6	0.9953	0.9955	0.9956	0.9957	0.9959	0.9960	0.9961	0.9962	0.9963	0.9964
...										
3.0	0.9987	0.9987	0.9987	0.9988	0.9988	0.9989	0.9989	0.9989	0.9990	0.9990

Commonly used z values:

AREA IN RIGHT TAIL, α	z_α
0.10	1.282
0.05	1.645
0.025	1.96
0.010	2.326
0.005	2.576

the probability of observing a z value between -1.96 and 1.96 is 95%. The reader should study the z table and associated graph of the z distribution to verify that the probabilities (or areas under the probability density function) described above are correct.

Often, we need to determine the probability that an experimental outcome falls between two values or that the outcome is greater than some value a or less or greater than some value b. To find these areas, we can use the following important formulas, where Pr is the probability:

$$\Pr(a \le z \le b) = \Pr(z \le b) - \Pr(z \le a)$$
$$= \text{area between } z = a \text{ and } z = b.$$

$$\Pr(z \le a) = 1 - \Pr(z < a)$$

$$= \text{area to right of } z = a$$
$$= \text{area in the right "tail".}$$

Thus, for any observation or measurement, x, from any normal distribution:

$$\Pr(a \le x \le b) = \Pr\left(\frac{a-\mu}{\sigma} \le z \le \frac{b-\mu}{\sigma}\right),$$

where μ is the mean of normal distribution and σ is the standard deviation of normal distribution.

In other words, we need to normalize or find the z values for each of our parameters, a and b, to find the area under the standard normal curve (z distribution) that represents the expression on the left side of the above equation.

Example 4.1 The mean intake of fat for males 6 to 9 years old is 28 g, with a standard deviation of 13.2 g. Assume that the intake is normally distributed. Steve's intake is 42 g and Ben's intake is 25 g.

1. What is the proportion of area between Steve's daily intake and Ben's daily intake?
2. If we were to randomly select a male between the ages of 6 and 9 years, what is the probability that his fat intake would be 50 g or more?

Solution: x = fat intake

1. The problem may be stated as: what is $\Pr(25 \leq x \leq 42)$?

   Assuming a normal distribution, we convert to z scores:

   What is $\Pr(((25 - 28)/13.2) < z < ((42 - 28)/13.2)))$?

   $= \Pr(-0.227 \leq z \leq 1.06) = \Pr(z \leq 1.06) - \Pr(z \leq -0.227)$ (using formula for $\Pr(a \leq z \leq b)$)

   $= \Pr(z \leq 1.06) - [1 - \Pr(z \leq 0.227)] = 0.8554 - [1 - 0.5910] = 0.4464$ or 44.6% of area under the z curve.

2. The problem may be stated as, "What is $\Pr(x > 50)$?"

   Normalizing to z score, what is $\Pr(z > (50 - 28)/13.2)$?

   $$= \Pr(z > 1.67)$$

   $$= 1 - \Pr(z \leq 1.67) = 1 - 0.9525 = 0.0475, \text{ or } 4.75\% \text{ of the area}$$

under the z curve.

Example 4.2 Suppose that the specifications on the range of displacement for a limb indentor are 0.5 ± 0.001 mm. If these displacements are normally distributed, with mean = 0.47 and standard deviation = 0.002, what percentage of indentors are within specifications?

Solution: x = displacement.

   The problem may be stated as, "What is $\Pr(0.499 \leq x \leq 0.501)$?"

   Using z scores, $\Pr(0.499 \leq x \leq 0.501) = \Pr((0.499 - 0.47)/0.002 \leq z \leq (0.501 - 0.47/0.002))$

   $$= \Pr(14.5 \leq z \leq 15.5) = \Pr(z \leq 15.5) - \Pr(z \leq 14.5) = 1 - 1 = 0$$

It is useful to note that if the distribution of the underlying population and the associated sample data are not normal (i.e. skewed), transformations may often be used to make the data normal, and the statistics covered in this text may then be used to perform statistical analysis on the transformed data. These transformations on the raw data include logs, square root, and reciprocal.

4.2 THE NORMAL DISTRIBUTION AND SAMPLE MEAN

All statistical analysis follows the same general procedure:

1. Assume an underlying distribution for the data and associated parameters (e.g., the sample mean).

2. Scale the data or parameter to a "standard" distribution.
3. Estimate confidence intervals using a standard table for the assumed distribution. (The question we ask is, "What is the probability of observing the experimental outcome by chance alone?")
4. Perform hypothesis test (e.g., Student's t test).

We are beginning with and focusing most of this text on the normal distribution or probability model because of its prevalence in the biomedical field and something called the central limit theorem. One of the most basic statistical tests we perform is a comparison of the means from two or more populations. The sample mean is it itself an estimate made from a finite number of samples. Thus, the sample mean, $\bar{x}$, is itself a random variable that is modeled with a normal distribution [4].

Is this model for $\bar{x}$ legitimate? The answer is yes, for large samples, because of the central limit theorem, which states [4, 10]:

If the individual data points or samples (each sample is a random variable), x_i, *come from any arbitrary probability distribution, the sum (and hence, average) of those data points is normally distributed as the sample size,* n, *becomes large.*

Thus, even if each sample, such as the toss of a dice, comes from a nonnormal distribution (e.g., a uniform distribution, such as the toss of a dice), the sum of those individual samples (such as the sum we use to estimate the sample mean, $\bar{x}$) will have a normal distribution. One can easily assume that many of the biological or physiological processes that we measure are the sum of a number of random processes with various probability distributions; thus, the assumption that our samples come from a normal distribution is not unreasonable.

4.3   CONFIDENCE INTERVAL FOR THE SAMPLE MEAN

Every sample statistic is in itself a random variable with some sort of probability distribution. Thus, when we use samples to estimate the true statistics of a population (which in practice are usually not known and not obtainable), we want to have some level of confidence that our sample estimates are close to the true population statistics or are representative of the underlying population or process.

In estimating a confidence interval for the sample mean, we are asking the question: "How close is our sample mean (estimated from a finite number of samples) to the true mean of the population?"

To assign a level of confidence to our statistical estimates or statistical conclusions, we need to first assume a probability distribution or model for our samples and underlying population and then we need to estimate a confidence interval using the assumed distribution.

The simplest confidence interval that we can begin with regarding our descriptive statistics is a confidence interval for the sample mean, $\bar{x}$. Our question is how close is the sample mean to the true population or process mean, μ?

Before we can answer this, we need to assume an underlying probability model for the sample mean, $\bar{x}$, and true mean, μ. As stated earlier, it may be shown that for a large samples size, the sample mean, $\bar{x}$, is well modeled by a normal distribution or probability model. Thus, we will use this model when estimating a confidence interval for our sample mean, $\bar{x}$.

Thus, $\bar{x}$ is estimated from the sample. We then ask, how close is $\bar{x}$ (sample mean) to the true population mean, μ?

It may be shown that if we took many groups of n samples and estimated $\bar{x}$ for each group,

1. the average or sample mean of $\bar{x} = \mu$, and
2. the standard deviation of $\bar{x} = s/\sqrt{n}$.

Thus, as our sample size, n, gets large, the distribution for $\bar{x}$ approaches a normal distribution.

For large n, $\bar{x}$ follows a normal distribution, and the z score for $\bar{x}$ may be used to estimate the following:

$$\Pr(a \le \bar{x} \le b) = \Pr\left(\frac{a - \mu}{\sigma/\sqrt{n}} \le z \le \frac{b - \mu}{\sigma/\sqrt{n}}\right).$$

This expression assumes a large n and that we know σ.

Now we look at the case where we might have a large n, but we do not know σ. In such cases, we replace σ with s to get the following expression:

$$\Pr(a \le \bar{x} \le b) = \Pr\left(\frac{a - \mu}{s/\sqrt{n}} \le z \le \frac{b - \mu}{s/\sqrt{n}}\right),$$

where $s/\sqrt{n}$ is called the sample standard error and represents the standard deviation for $\bar{x}$.

Let us assume now for large n, we want to estimate the 95% confidence interval for $\bar{x}$. We first scale the sample mean, $\bar{x}$, to a z value (because the central limit theorem says that $\bar{x}$ is normally distributed)

$$z = \frac{\bar{x} - \mu}{s/\sqrt{n}}.$$

We recall that 95% of z values fall between ±1.96 (approximately 2σ) of the mean, and for the z distribution,

$$\Pr(-1.96 \leq z \leq 1.96) = 0.95.$$

Substituting for z,

$$z = \frac{\bar{x} - \mu}{s/\sqrt{n}}.$$

we get

$$0.95 = \Pr\left(-1.96 \leq \frac{\bar{x} - \mu}{s/\sqrt{n}} \leq 1.96\right).$$

If we use the following notation in terms of the sample standard error:

$$SE(\bar{x}) = \frac{s}{\sqrt{n}}.$$

Rearranging terms for the expression above, we note that the probability that μ lies between ±1.96 (or 2) standard deviations of $\bar{x}$ is 95%:

$$0.95 = \Pr\left(\bar{x} - 1.96 SE(\bar{x}) \leq \mu \leq \bar{x} + 1.96 SE(\bar{x}).\right)$$

Note that 1.96 is referred to as $z_{\alpha/2}$. This z value is the value of z for which the area in the right tail of the normal distribution is $\alpha/2$. If we were to estimate the 99% confidence interval, we would substitute $z_{0.01/2}$, which is 2.576, into the 1.96 position above.

Thus, For large n and any confidence level, $1 - \alpha$, the $1 - \alpha$ confidence interval for the true population mean, μ, is given by:

$$\mu = \bar{x} \pm z_{\alpha/2} SE(\bar{x}).$$

This means that there is a $(1 - \alpha)$percent probability that the true mean lies within the above interval centered about $\bar{x}$.

Example 4.3 Estimate of confidence intervals
Problem: Given a collection of data with, $\bar{x} = 505$ and $s = 100$. If the number of samples was 1000, what is the 95% confidence interval for the population mean, μ?

Solution: If we assume a large sample size, we may use the z distribution to estimate the confidence interval for the sample mean using the following equation:

$$\mu = \bar{x} \pm z_{\alpha/2} \, \text{SE}(\bar{x}).$$

We plug in the following values:

$$\bar{x} = 505;$$
$$\text{SE}(\bar{x}) = s / \sqrt{n} = 100 / \sqrt{1000}.$$

For 95% confidence interval, $\alpha = 0.05$.

Using a z table to locate $z(0.05/2)$, we find that the value of z that gives an area of 0.025 in the right tail is 1.96

Plugging in $\bar{x}$, $\text{SE}(\bar{x})$, and $z(\alpha/2)$ into the estimate for the confidence interval above, we find that the 95% confidence interval for $\mu = [498.80, 511.20]$.

Note that if we wanted to estimate the 99% confidence interval, we would simply use a different z value, $z(0.01/2)$ in the same equation. The z value associated with an area of 0.005 in the right tail is 2.576. If we use this z value, we estimate a confidence interval for μ of $[496.86, 515.14]$. We note that the confidence interval has widened as we increased our confidence level.

4.4   THE t DISTRIBUTION

For small samples, $\bar{x}$ is no longer normally distributed. Therefore, we use Student's t distribution to estimate the true statistics of the population. The t distribution, as illustrated in Table 4.2 looks like a z distribution but with slower taper at the tails and flatter central region.

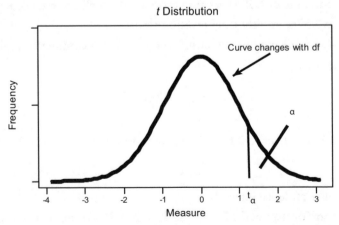

Table entry = $t(\alpha; df)$, where α is the area in the tail to the right of $t(\alpha; df)$ and df is degrees of freedom.

TABLE 4.2: Percentage points for Student's t distribution

df	α = area to right of $t(\alpha; df)$				
	0.10	0.05	0.025	0.01	0.005
1	3.078	6.314	12.706	31.821	63.657
2	1.886	2.920	4.303	6.965	9.925
3	1.638	2.353	3.182	4.541	5.841
4	1.533	2.132	2.776	3.747	4.604
5	1.476	2.015	2.571	3.365	4.032
...					
10	1.372	1.812	2.228	2.764	3.169
11	1.363	1.796	2.201	2.718	3.106
12	1.356	1.782	2.179	2.681	3.055
13	1.350	1.771	2.160	2.650	3.012
14	1.345	1.761	2.145	2.624	2.977
15	1.341	1.753	2.131	2.602	2.947
...					
30	1.310	1.697	2.042	2.457	2.750
40	1.303	1.684	2.021	2.423	2.704
60	1.296	1.671	2.000	2.390	2.660
120	1.289	1.658	1.980	2.358	2.617
∞	1.282	1.645	1.960	2.326	2.576
z_α(large sample)	1.282	1.645	1.960	2.326	2.576

We use t distribution, with slower tapered tails, because with so few samples, we have less certainty about our underlying distribution (probability model).

Now our normalized value for $\bar{x}$ is given by

$$\frac{\bar{x} - \mu}{s/\sqrt{n}},$$

which is known to have a t distribution rather than the z distribution that we have discussed thus far. The t distribution was first invented by W.S. Gosset [4, 11], a chemist who worked for a brewery. Gosset decided to publish his t distribution under the alias of Student. Hence, we often refer to this distribution as Student's t distribution.

The t distribution is symmetric like the z distribution and generally has a bell shape. But the amount of spread of the distribution to the tails, or the width of the bell, depends on the sample size, n. Unlike the z distribution, which assumes an infinite sample size, the t distribution changes shape with sample size. The result is that the confidence intervals estimated with t values are more "spread out" than for z distribution, especially for small sample sizes, because with such samples sizes, we are penalized for not having sufficient samples to represent the extent of variability of the underlying population or process. Thus, when we are estimating confidence intervals for the sample mean, $\bar{x}$, we do not have as much confidence in our estimate. Thus, the interval widens to reflect this decreased certainty with smaller sample sizes. In the next section, we estimate the confidence interval for the same example given previously, but using the t distribution instead of the z distribution.

4.5   CONFIDENCE INTERVAL USING t DISTRIBUTION

Like the z tables, there are t tables where the values of t that are associated with different areas under the probability curve are already calculated and may be used for statistical analysis without the need to recalculate the t values. The difference between the z table and t table is that now the t values are a function of the samples size or degrees of freedom. Table 4.2 gives a few lines of the t table from [3].

To use the table, one simply looks for the intersection of degrees of freedom, df, (related to sample size) and α value that one desires in the right-hand tail. The intersection provides the t value for which the area under the t curve in the right tail is α. In other words, the probability that t will be less than or equal to a specific entry in the table is $1 - \alpha$. For a specific sample size, n, the degrees of freedom, $df = n - 1$.

Now we simply substitute t for z to find our confidence intervals. So, the confidence interval for the sample mean, $\bar{x}$, using the t distribution now becomes

$$\mu = \bar{x} \pm t\left(\frac{\alpha}{2}; n-1\right) SE(\bar{x})$$

where μ is the true mean of the underlying population or process from which we are drawing samples, $SE(\bar{x})$ is the standard error of the underlying population or process, t is the t value for which there is an area of $\alpha/2$ in the right tail, and n is the sample size.

Example 4.4 Confidence interval using t distribution

Problem: We consider the same example used previously for estimating confidence intervals using z values. In this case, the sample size is small ($n = 20$), so we now use a t distribution.

Solution: Now, our estimate for confidence interval for $\mu = \bar{x} \pm t\left(\frac{\alpha}{2}; n-1\right) SE(\bar{x})$.

Again, we plug in the following values:

$$\bar{x} = 505;$$
$$SE(\bar{x}) = s/\sqrt{n} = 100/\sqrt{20}.$$

For 95% confidence interval, $\alpha = 0.05$.

Using a t table to locate $t(0.05/2, 20 - 1)$, we find that for 19 df, the value of t that gives an area of 0.025 in the right tail is 2.093.

Plugging in $\bar{x}$, $SE(\bar{x})$, and $t(\alpha, n - 1)$ into the estimate for the confidence interval above, we find that the 95% confidence interval for $\mu = [458.20, 551.80]$ We note that this confidence interval is widened compared with that previously estimated using the z values. This is expected because the t distribution is wider than the z distribution at small sample sizes, reflecting the fact that we have less confidence in our estimate of $\bar{x}$ and, hence, μ when the sample size is small.

Confidence intervals may be estimated for most descriptive statistics, such as the sample mean, sample variance, and even the line of best fit determined through linear regression [3]. As noted above, the confidence interval reflects the variability in the parameter estimate and is influenced by the sample size, the variability of the population, and the level on confidence that we desire. The greater the desired confidence, the wider the interval. Likewise the greater the variability in the samples, the wider the confidence interval.

Example 4.5 Review of probability concepts

Problem: Assembly times were measured for a sample of 15 glucose infusion pumps. The mean time to assemble a glucose infusion pump was 15.8 minutes, with a standard deviation of 2.4 minutes. Assuming a relatively symmetric distribution for assembly times,

1. What percentage of infusion pumps require more than 17 seconds to assemble?
2. What is the 99% confidence interval for the true mean assembly time (μ)?
3. What is the 99% confidence interval for mean assembly time if the sample size is 2500?

Solution:

1. x = assembly time

    What is the $\Pr(x > 17)$?

    $\Pr(x > 17) = \Pr(z > (17 - 15.8)/2.4) = \Pr(z > 0.5) = 1 - \Pr(z \leq 0.5) = 1 - 0.6915 = 0.3085$, or 38.05% of the infusion pumps.

2. $= \bar{x} \pm t(\alpha/2, n - 1)SE(x)$

    $= 15.8 \pm t((0.01)/2; 15 - 1)\,(2.4\sqrt{15})$

    $= 15.8 \pm t(0.005, 14)\,(0.6196)$

    $= 15.8 \pm 2.977\,(0.6196)$

    $= [13.96, 17.64]$

    Because the samples size of 2500 is now large, we use a z value for estimating the confidence interval, the $\mu = \bar{x} \pm z\,(\alpha/2)SE(x)$

3. $= \bar{x} \pm z\,(\alpha/2)SE(x)$

    $= 15.8 \pm z\,(0.01/2)\,(2.4\sqrt{2500})$

    $= 15.8 \pm 2.576\,(0.048)$

    $= [15.68, 15.92]$

· · · ·

CHAPTER 5

Statistical Inference

Now that we have collected our data, estimated some basic descriptive statistics and assumed a probability model for the underlying population or process, we are prepared to perform some statistical analysis that will allow us to compare populations and test hypotheses. For this text, we are only going to discuss statistical analysis for which the normal distribution is a good probability model for the underlying population(s). If the data or samples suggest that the underlying population or process is not well modeled by a normal distribution, then we will need to resort to other types of statistical analysis, such a nonparametric techniques [6], which do not assume an underlying probability model. Some of these tests include the Wilcoxon rank sum test, Mann–Whitney U test, Kruskal–Wallis test, and the runs test [6, 12]. More advanced texts provide details for administering these nonparametric tests. Keep in mind that if the statistical analysis presented in this text is used for populations or processes that are not normally distributed, the results may be of little value, and the investigator may miss important findings from the data.

Assuming our data represents a normally distributed population or process, we are now prepared to perform a variety of statistical tests that allow us to test hypotheses about the equality of means and variances across two or more populations. Remember that for normal distributions, only the mean and standard deviation are required to completely characterize the probabilistic nature of the population or process. Thus, if we are to compare two normally distributed populations, we need only compare the means and variances of the populations. If the populations or processes are not normally distributed, there may be other parameters, such as skew and kurtosis, which differentiate two or more populations or processes.

5.1 COMPARISON OF POPULATION MEANS

One of the most fundamental questions asked by scientists or engineers performing experiments is whether two populations, methods, or treatments are really different in central tendency. More specifically, are the two population means, reflected in the sample means collected under two different experimental conditions, significantly different? Or, is the observed difference between two means simply because of chance alone?

Some examples or questions asked by biomedical engineers that require comparing two population means include the following:

1. Is one chemotherapy drug more effective than a second chemotherapy drug in reducing the size of a cancerous tumor?
2. Is there a difference in postural stability between young and elderly populations?
3. Is there a difference in bone density for women who are premenopausal versus those who are postmenopausal?
4. Is titanium a stronger material for bone implants than steel?
5. For MRI, does one type of pulse sequence perform better than another in detecting white matter tracts in the brain?
6. Do drug-eluting intravascular stents prevent restenosis more effectively than noncoated stents?

In answering these type of questions, biomedical engineers often collect samples from two different groups of subjects or from one group of subjects but under two different conditions. For a specific measure that represents the samples, a sample mean is typically estimated for each group or under each of two conditions. Comparing the means from two populations reflected in two sets of data is probably the most reported statistical analysis in the scientific and engineering literature and may be accomplished using something known as a *t* test.

5.1.1   The *t* Test

When performing the *t* test, we are asking the question, "Are the means of two populations really different?" or "Would we see the observed differences simply because of random chance?" The two populations are reflected in data that have been collected under two different conditions. These conditions may include two different treatments or processes.

To address this question, we use one of the following two tests, depending on whether the two data sets are independent or dependent on one another:

1. Unpaired *t* test for two sets of independent data
2. Paired *t* test for two sets of dependent data

Before we describe each type of *t* test, we need to discuss the notion of hypothesis testing.

5.1.1.1   Hypothesis Testing

Whenever we perform statistical analysis, we are testing some hypothesis. In fact, before we even collect our data, we formulate a hypothesis and then carefully design our experiment and collect and analyze the data to test the hypothesis. The outcome of the statistical test will allow us, if the assumed probability model is valid, to accept or reject the hypothesis and do so with some level of confidence.

There are basically two forms of hypotheses that we test when performing statistical analysis. There is the null hypothesis and the alternative hypothesis.

The null hypothesis, denoted as H_0, is expressed as follows for the t test comparing two population means, μ_1 and μ_2:

$$H_0: \mu_1 = \mu_2.$$

The alternative hypothesis, denoted as H_1, is expressed as one of the following for the t test comparing two population means, μ_1 and μ_2:

$$H_1: \mu_1 \neq \mu_2 \qquad \text{(two-tailed t test),}$$

$$H_1: \mu_1 < \mu_2 \qquad \text{(one tailed t test),}$$

or

$$H_1: \mu_1 > \mu_2 \qquad \text{(one-tailed t test).}$$

If the H_1 is of the first form, where we do not know in advance of the data collection which mean will be greater than the other, we will perform a two-tailed test, which simply means that the level of significance for which we accept or reject the null hypothesis will be double that of the case in which we predict in advance of the experiment that one mean will be lower than the second based on the physiology or the engineering process or other conditions affecting experimental outcome. Another way to express this is that with a one-tailed test, we will have greater confidence in rejecting or accepting the null hypothesis than with the two-tailed condition.

5.1.1.2 Applying the t Test

Now that we have stated our hypotheses, we are prepared to perform the t test. Given two populations with n_1 and n_2 samples, we may compare the two population means using a t test. It is important to remember the underlying assumptions made in using the t test to compare two population means:

1. The underlying distributions for both populations are normal.
2. The variances of the two populations are approximately equal: $s_1^2 = s_2^2$.

These are big assumptions we make when n_1 and n_2 are small. If these assumptions are poor for the data being analyzed, we need to find different statistics to compare the two populations.

Given two sets of sampled data, x_i and y_i, the means for the two populations or processes reflected in the sampled data can be compared using the following t statistic:

5.1.1.3 Unpaired t Test

The unpaired t statistic may be estimated using:

$$T = \frac{\bar{x} - \bar{y}}{\sqrt{\dfrac{(n_1 - 1)S_x^2 + (n_2 - 1)S_y^2}{n_1 + n_2 - 2}\left(\dfrac{1}{n_1} + \dfrac{1}{n_2}\right)}},$$

where n_1 is the number of x_i observations, n_2 is the number of y_i observations, S_x^2 is the sample variance of x_i, S_y^2 = sample variance of y_i, $\bar{x}$ is the sample average for x_i, and $\bar{y}$ is the sample average for y_i.

Once the T statistic has been computed, we can compare our estimated T value to t values given in a table for the t distribution. More specifically, if we want to reject the null hypothesis with $1 - \alpha$ level of confidence, then we need to determine whether our estimated T value is greater than the t value entry in the t table associated with a significance level of α (one-sided t test) or $\alpha/2$ (two-sided t test). In other words, we need to know the t value from the t distribution for which the area under the t curve to the right of t is α. Our estimated T must be greater than this t to reject the null hypothesis with $1 - \alpha$ level of confidence.

Thus, we compare our T value to the t distribution table entry for

$$t(\alpha, n_1 + n_2 - 2) \qquad \text{(one-sided)}$$

or

$$t(\alpha/2, n_1 + n_2 - 2) \qquad \text{(two-sided)},$$

where α is the level of significance (equal to $1 -$ level of confidence), and n_1 and n_2 are the number of samples from each of the two populations being compared.

Note that α is the level of significance at which we want to accept or reject our hypothesis. For most research, we reject the null hypothesis when $\alpha \leq 0.05$. This corresponds to a confidence level of 95%.

For example, if we want to reject our null hypothesis with a confidence level of 95%, then for a one-sided t test, our estimated T must be greater than $t(0.05, n_1 + n_2 - 2)$ to reject H_0 and accept H_1 with 95% confidence or a significance level of 0.05 or less. Remember that our confidence = $(1 - 0.05) \times 100\%$. This test was for $H_1: \mu_1 < \mu_2$ or $\mu_1 > \mu_2$ (one-sided). If our H_1 was H_1:

$\mu_1 \neq \mu_2$, then our measured T must be greater than $t(0.05/2, n_1 + n_2 - 2)$ to reject the null hypothesis with the same 95% confidence.

Thus, to test the following alternative hypotheses:

1. For $H_1: \mu_1 \neq \mu_2$: Use two-tail t test. For $(\alpha < 0.05)$, the estimated $T > t(0.025, n_1 + n_2 - 2)$ to reject H_0 with 95% confidence.
2. For $H_1: \mu_1 < \mu_2$ or $H_1: \mu_1 < \mu_2$: Use one-tail t test. For $(\alpha < 0.05)$, estimated $T > t(0.05, n_1 + n_2 - 2)$ to reject H_0 with 95% confidence.

α is also referred to as the type I error. For the t test, α is the probability of observing a measured T value greater than the table entry, $t(\alpha; df)$ if the true means of two underlying populations, x and y, were actually equal. In other words, there is no significant $(\alpha < 0.05)$ difference in the two population means, but the sampled data analysis led us to conclude that there is a difference and thus reject the null hypothesis. This is referred to as a type I error.

Example 5.1 An example of a biomedical engineering challenge where we might use the t test to improve the function of medical devices is in the development of implantable defibrillators for the detection and treatment of abnormal heart rhythms [13]. Implantable defibrillators are small electronic devices that are placed in the chest and have thin wires that are placed in the chambers of the heart. These wires have electrodes that detect the small electrical current traversing the heart muscle. Under normal conditions, these currents follow a very orderly pattern of conduction through the heart muscle. An example of an electrogram, or record of electrical activity that is sensed by an electrode, under normal conditions is given in Figure 5.1. When the heart does not function normally and enters a state of fibrillation (Figure 5.1), whereby the heart no longer contracts normally or pumps blood to the rest of the body, the device should shock the heart with a large electrical current from the device in an attempt to convert the heart rhythm back to normal. For a number of reasons, it is important that the device accurately detect the onset of life-threatening arrhythmias, such as fibrillation, and administer an appropriate shock. To administer a shock, the device must use some sort of signal processing algorithms to automatically determine that the electrogram is abnormal and characteristic of fibrillation.

One algorithm that is used in most devices for differentiating normal heart rhythms from fibrillation is a rate algorithm. This algorithm is basically an amplitude threshold crossing algorithm whereby the device determines how often the electrogram exceeds an amplitude threshold in a specified period of time and then estimates a rate from the detected threshold crossings. Figure 5.2 illustrates how this algorithm works.

Before such an algorithm was put into implantable defibrillators, researchers and developers had to demonstrate whether the rate estimated by such a device truly differed between normal and

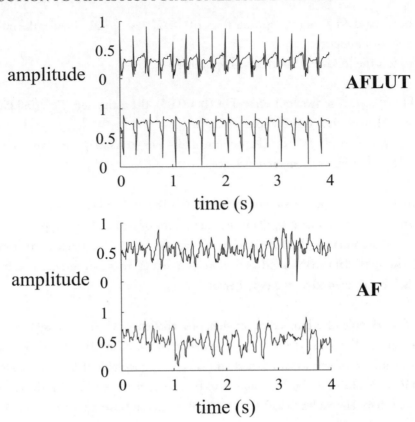

FIGURE 5.1: Electrogram recordings measured from electrodes placed inside the left atrium of the heart. For each rhythm, there are two electrogram recordings taken from two different locations in the atrium: atrial flutter (AFLUT) and atrial fibrillation (AF).

fibrillatory rhythms. It was important to have little overlap between the normal and fibrillatory rhythms so that a rate threshold could be established, whereby rates that exceeded the threshold would lead to the device administering a shock. For this algorithm to work, there must be a significant difference in rates between normal and fibrillatory rhythms and little overlap that would lead to shocks being administered inappropriately and causing great pain or risk the induction of fibrillation. Overlap in rates between normal and fibrillatory rhythms could also result in the device missing detection of fibrillation because of low rates.

To determine if rate is a good algorithm for detecting fibrillatory rhythms, investigators might actually record electrogram signals from actual patients who have demonstrated normal or fibrillatory rhythms and use the rate algorithm to estimate a rate and then compare the mean rates for normal rhythms against mean rates for fibrillatory rhythms. For example, let us assume

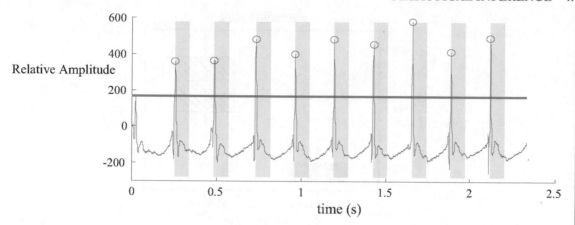

FIGURE 5.2: A heart rate is estimated from an electrogram using an amplitude threshold algorithm. Whenever the amplitude of the electrogram signal (solid waveform) exceeds an amplitude threshold (solid gray horizontal line) within a specific time interval (shaded vertical bars), an "event" is detected. A rate is calculated by counting the number of events in a certain time period.

investigators collected 15-second electrogram recordings for examples of fibrillatory (n_1 = 10) and nonfibrillatory rhythms (n_2 = 11) in 21 patients. We note that the fibrillatory data were collected from different subjects than the normal data. Thus, we do not have blocking in this experimental design.

A rate was estimated, using the device algorithm, for each individual 15-second recording. Figure 5.3 shows a box-and-whisker plot for each of the two data sets.

The descriptive statistics for the two data sets are given by the following table:

	NONFIBRILLATORY ELECTROGRAMS (N_1 = 11)	FIBRILLATORY ELECTROGRAMS (N_2 = 10)
Mean	96.82	239.0
Standard deviation	22.25	55.3

There are several things to note from the plots and descriptive statistics. First, the sample size is small; thus, it is likely that the variability of electrogram recordings from normal and fibrillatory rhythms is not adequately represented. Moreover, the data are skewed and do not appear to be well modeled by a normal distribution. Finally, the variability in rates appears to be much greater for fibrillatory rhythms than for normal rhythms. Thus, some of the assumptions we make regarding

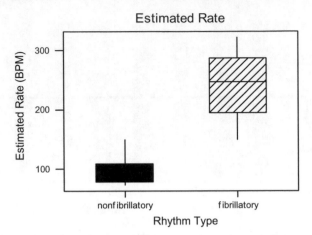

FIGURE 5.3: Box-and-whisker plot for rate data estimated from samples of nonfibrillatory and fibrillatory heart rhythms.

the normality of the populations and the equality of variances are probably violated in applying the t test. However for the purpose of illustration, we will perform a t test using the unpaired t statistic.

To find an estimated T statistic from our sample data, we plug in the means, variances, and sample sizes into the equation for the unpaired T statistic. We turn the crank and find that for this example, the estimated t value is 7.59. If we look at our t tables for [11 + 10 − 2] degrees of freedom, we can reject H_0 at $\alpha < 0.005$ because our estimated T value exceeds the t table entries for $\alpha = 0.05$ ($t = 1.729$), $\alpha = 0.01$ ($t = 2.539$), and $\alpha = 0.005$ ($t = 2.861$).

Thus, we reject the null hypothesis with $(1 − 0.005) \times 100\%$ confidence and state that mean heart rate for fibrillatory rhythms is greater than mean heart rate for normal rhythms. Thus, the rate algorithm should perform fairly well in differentiating normal from fibrillatory rhythms. However, we have only tested the population means. As one may see in Figure 5.3, there is some overlap in individual samples between normal and fibrillatory rhythms. Thus, we might expect the device to make errors in administering shock inappropriately when the heart is in normal but accelerated rhythms (as in exercise), or the device may fail to shock when the heart is fibrillating but at a slow rate or with low amplitude.

In applying the t test, it is important to note that you can never prove that two means are equal. The statistics can only show that a specific test cannot find a difference in the population means, and, not finding a difference is not the same as proving equality. The null or default hypothesis is that there is no difference in the means, regardless of what the true difference is between the two means. Not finding a difference with the collected data and appropriate statistical test does not mean that the means are proved equal. Thus, we do not accept the null hypothesis with a level of significance. We simply accept the null hypothesis and do not accept with a confidence or significance level. We only assign a level of confidence when we reject the null hypothesis.

5.1.1.4   Paired *t* Test

In the previous example, we used an unpaired *t* test because the two data sets came from two different, unrelated, groups of patients. The problem with such an experimental design is that differences in the two patient populations may lead to differences in mean heart rate that have nothing to do with the actual heart rhythm but rather differences in the size or age of the hearts between the two groups of patients or some other difference between patient groups. A better way to conduct the previous study is to collect our normal and fibrillatory heart data from one group of patients. A single defibrillator will only need to differentiate normal and fibrillatory rhythms within a single patient. By blocking on subjects we can eliminate the intersubject variability in electrogram characteristics that may plague the rate algorithm from separating populations. It may be more reasonable to assume that rates for normal and fibrillatory rhythms differ more within a patient than across patients. In other words, for each patient, we collect an electrogram during normal heart function and fibrillation. In such experimental design, we would compare the means of the two data sets using a paired *t* test.

We use the paired *t* test when there is a natural pairing between each observation in data set, *X*, with each observation in data set, *Y*. For example, we might have the following scenarios that warrant a paired *t* test:

1. Collect blood pressure measurements from 8 patients before and after exercise.
2. Collect both MR and CT data from the same 20 patients to compare quality of blood vessel images.
3. Collect computing time for an image processing algorithm before and after a change is made in the software.

In such cases, the *X* and *Y* data sets are no longer independent. In the first example above, because we are collecting the before and after data from the same experimental units, the physiology and biological environment that impacts blood pressure before exercise in each patients also affects blood pressure after exercise. Thus, there are a number of variables that impact blood pressure that we cannot directly control but whose effects on blood pressure (beside the exercise effect) can be controlled by using the same experimental units (human subjects in this case) for each data set. In such cases, the experimental units serve as their own control. This is typically the preferred experimental design for comparing means.

For the paired t test, we again have a null hypothesis and an alternative hypothesis as stated above for the unpaired t test. However, in a paired t test, we use a t test on the difference, $W_i = X_i - Y_i$, between the paired data points from each of the two populations.

We now calculate the paired T statistic: (n = number of pairs)

$$T = \frac{\overline{W}}{S_w / \sqrt{n}},$$

where $\overline{W}$ is the average difference of the differences, W_i, and S_w is the standard deviation of the differences, W_i.

As for the unpaired t test, we now have an estimated T value that we can compare with the t values in the table for the t distribution to determine if the estimated T value lies in the extreme values (greater than 95%) of the t distribution.

To reject the null hypothesis at a significance level of α (confidence level of $1 - \alpha$), our estimated T value must be greater than $t\,(\alpha, n - 1)$, where n is the number of pairs of data.

If the estimated T value exceeds $t\,(\alpha, n - 1)$, we reject H_0 and accept H_1 at a significance level of α or a confidence level of $1 - \alpha$.

Once again, if H_1: $\mu_1 < \mu_2$ or H_1: $\mu_1 < \mu_2$, we perform a one-sided test where our T statistic must be greater than $t\,(\alpha, n - 1)$ to reject the null hypothesis at a level of α. If the null hypothesis that we begin with before the experiment is H_1: ($\mu_1 \neq \mu_2$), then we perform a two-sided test, and the T statistic must be greater than $t\,(\alpha/2, n - 1)$ to reject the null hypothesis at a significance level of α.

5.1.1.5   Example of a Biomedical Engineering Challenge

In relation to abnormal heart rhythms discussed previously, antiarrhythmic drugs may be used to slow or terminate an abnormal rhythm, such as fibrillation. For example, a drug such as procainamide may be used to terminate atrial fibrillation. The exact mechanism whereby the drug leads to termination of the rhythm is not exactly known, but it is thought that the drug changes the refractory period and conduction velocity of the heart cells [8]. Biomedical engineers often use signal processing on the electrical signals generated by the heart as an indirect means for studying the underlying physiology. More specifically, engineers may use spectral analysis or Fourier analysis to look at changes in the spectral content of the electrical signal, such as the electrogram, over time. Such changes may tell us something about the underlying electrophysiology.

For example, spectral analysis has been used to look at changes in the frequency spectrum of atrial fibrillation with drug administration. In one such study [8], biomedical engineers were interested in looking at changes in median frequency of atrial electrograms after drug administration. Figure 5.4 shows an example of the frequency spectrum for atrial fibrillation and the location

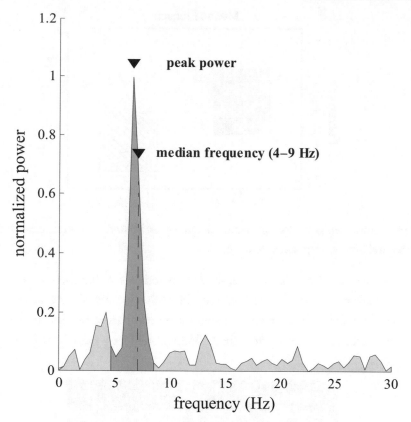

FIGURE 5.4: Frequency spectrum for an example of atrial fibrillation. Median frequency is defined as that frequency that divides the area under the power curve in the 4- to 9-Hz band in half.

of the median frequency, which is the frequency that divides the power of the spectrum (area under the spectral curve between 4 and 9 Hz) in half. One question posed by the investigators is whether median frequency decreases after administration of a drug such procainamide, which is thought to slow the electrical activity of the heart cells.

Thus, an experiment was conducted to determine whether there was a significant difference in mean median frequency between fibrillation before drug administration and fibrillation after drug administration. Electrograms were collected in the right atrium in 11 patients before and after the drug was administered. Fifteen-second recordings were evaluated for the frequency spectrum, and the median frequency in the 4- to 9-Hz frequency band was estimated before and after drug administration.

Figure 5.5 illustrates the summary statistics for median frequency before and after drug administration. The question is whether there was a significant decrease in median frequency after

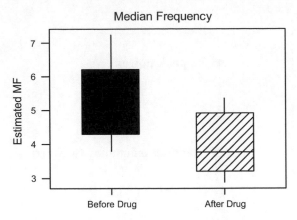

FIGURE 5.5: Box-and-whisker plot for median frequency estimated from samples of atrial fibrillation recorded before, and after a drug was administered.

administration of procainamide. In this case, the null hypothesis is that there is no change in mean median frequency after drug administration. The alternative hypothesis is that mean median frequency decreases after drug administration. Thus, we are comparing two means, and the two data sets have been collected from one set of patients. We will require a paired t test to reject or accept the null hypothesis.

BEFORE DRUG	AFTER DRUG	WI
4.30	2.90	1.4
4.15	2.97	1.18
3.80	3.20	0.60
5.10	3.30	1.80
4.30	3.75	0.55
7.20	5.35	1.85
6.40	5.10	1.30
6.20	4.90	1.30
6.10	4.80	1.30
5.00	3.70	1.30
5.80	4.50	1.30

To perform a paired t test, we create another column, W_i, as noted in the table above. We find the following for W_i: $\overline{W}$ = 1.262 and S_w = 0.402. If we use these estimates for $\overline{W}$ and S_w in our equation for the paired T statistic, we find that T = 10.42 for n = 11 pairs of data. If we compare our estimated T value to the t distribution, we find that our T value is greater than the table entry for t (0.005, 11 − 1);therefore, we may reject H_0 at a significance less than 0.005. In other words, we reject the null hypothesis at the $[1 − 0.005] \times 100\%$ confidence level.

Errors in Drawing Conclusions From Statistical Tests

When we perform statistical analysis, such as the t test, there is a chance that we are mistaken in rejecting or accepting the null hypothesis. When we draw conclusions from a statistical analysis, we typically assign a confidence level to our conclusion. This means that there is always some chance that our conclusions are incorrect.

There are two types of errors that may occur when we draw conclusions from a statistical analysis. These errors are referred to as types I and II.

Type I errors:

1. also referred to as a false-positive error;
2. occurs when we accept H_1 when H_0 is true;
3. may result in a false diagnosis of disease.

Type II errors:

1. also referred to as a false-negative error;
2. occurs when we accept H_0 when H_1 is true;
3. may result in a missed diagnosis (often more serious than a type I error).

If we think about the medical environment, a type I error might occur when a person is given a diagnostic test to detect streptococcus bacteria, and the test indicates that the person has the streptococcus bacteria when in fact, the person does not have the bacteria. The result of such an error means that the person spends money for an antibiotic that is serving no purpose.

A type II error would occur when the same person actually has the streptococcus bacteria, but the diagnostic test results in a negative outcome, and we conclude that the person does not have the streptococcus bacteria. In this example, this type of error is more serious than the type I error because the streptococcus bacteria left untreated can lead to a number of serious complications for the body.

The α value that we refer to as the level of significance is also the probability of a type I error. Thus, the smaller the level of significance at which we may reject the null hypothesis, the smaller the type I error and the lower the probability of making a type I error.

We typically use β to denote the type II error. We will discuss this error further at the end of chapter seven when we discuss power tests.

5.2   COMPARISON OF TWO VARIANCES

We used the t test to compare the means for two populations or processes. Two populations may also be compared for differences in variance. As discussed earlier, populations that are normally distributed are completely characterized by their mean and variance. Thus, if we want to test for differences between two normal populations, we need only compare their two means and their two variances.

Figure 5.6 illustrates the probability density functions for two normal populations (black and red traces). The four diagrams illustrate how two different normally distributed populations may compare with each other. The right two panels differ from the left two panels in the means of the populations. The top two panels differ from the bottom two panels in variance of the populations.

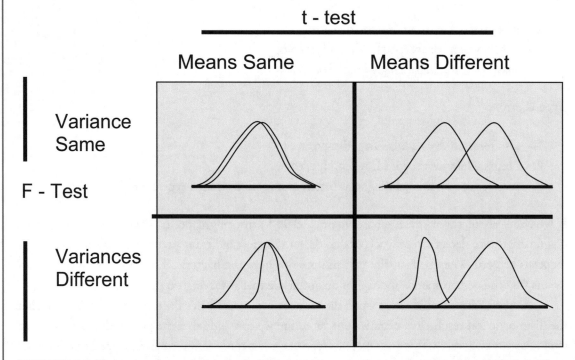

FIGURE 5.6: Two normal populations may differ in their means (top row), their variances (left half), or both (bottom right corner). t and F tests may be used to test for significant differences in the population means and population variances, respectively.

As indicated across the top of the tracings, a t test is used to test for differences in mean between the two populations. As indicated along the vertical direction, an F test is used to test for significant differences in the variances of the populations. Note that two normal populations may differ significantly in both mean and variance.

To compare the variances of two populations, we use what is referred to as an F test. As for the t test, the F test assumes that the data consist of independent random samples from each of two normal populations. If the two populations are not normally distributed, the results of the F test may be meaningless.

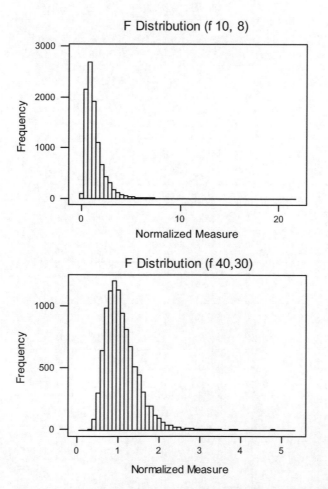

FIGURE 5.7: Histograms of samples drawn from two different F distributions. In the top panel, the two degrees of freedom are 10 and 8. In the lower panel, the two degrees of freedom are 40 and 30.

TABLE 5.1: Values from the F distribution for areas of α in the tail to the right of $F(dn, dd, \alpha)$

| dd | \multicolumn{7}{c}{dn} |
	1	2	3	4	5	...	10	11
1	161, *4052*	200, *4999*	216, *5403*	225, *5625*	230, *5764*		242, *6056*	243, *6082*
2	18.51, *98.49*	19.00, *99.01*	19.16, *99.17*	19.25, *99.25*	19.30, *99.30*		19.39, *99.40*	19.40, *99.41*
3	10.13, *34.12*	9.55, *30.81*	9.28, *29.46*	9.12, *28.71*	9.01, *28.24*		8.78, *27.23*	8.76, *27.13*
4	7.71, *21.20*	6.94, *18.00*	6.59, *16.69*	6.39, *15.98*	6.26, *15.52*		5.96, *14.54*	5.93, *14.45*
5	6.61, *16.26*	5.79, *13.27*	5.41, *12.06*	5.19, *11.39*	5.05, *10.97*		4.74, *10.05*	4.70, *9.96*
⋮								
10	4.96, *10.04*	4.10, *7.56*	3.71, *6.55*	3.48, *5.99*	3.33, *5.64*		2.97, *4.85*	2.94, *4.78*
12	4.75, *9.33*	3.88, *6.93*	3.49, *5.95*	3.26, *5.41*	3.11, *5.06*		2.76, *4.30*	2.72, *4.22*
15	4.54, *8.68*	3.68, *6.36*	3.29, *5.42*	3.06, *4.89*	2.90, *4.56*		2.55, *3.80*	2.51, *3.73*
⋮								
20	4.35, *8.10*	3.49, *5.85*	3.10, *4.94*	2.87, *4.43*	2.71, *4.10*		2.35, *3.37*	2.31, *3.30*
50	4.03, *7.17*	3.18, *5.06*	2.79, *4.20*	2.56, *3.72*	2.40, *3.41*		2.02, *2.70*	1.98, *2.62*
100	3.94, *6.90*	3.09, *4.82*	2.70, *3.98*	2.46, *3.51*	2.30, *3.20*		1.92, *2.51*	1.88, *2.43*
200	3.89, *6.76*	3.04, *4.71*	2.65, *3.38*	2.41, *3.41*	2.26, *3.11*		1.87, *2.41*	1.83, *2.34*
∞	3.84, *6.64*	2.99, *4.60*	2.60, *3.78*	2.37, *3.32*	2.21, *3.02*		1.83, *2.32*	1.79, *2.24*

dn = degrees of freedom for numerator; dd = degrees of freedom for denominator; α = the area in distribution tail to right of $F(dn, dd, \alpha)$ = 0.05 or **0.01**.

dn									
12	14	...	20	30	40	50	100	200	∞
244, *6106*	245, *6142*		248, *6208*	250, *6258*	251, *6286*	252, *6302*	253, *6334*	254, *6352*	254, *6366*
19.41, *99.42*	19.42, *99.43*		19.44, *99.45*	19.46, *99.47*	19.47, *99.48*	19.47, *99.48*	19.49, *99.49*	19.49, *99.49*	19.50, *99.50*
8.74, *27.05*	8.71, *26.92*		8.66, *26.69*	8.62, *26.50*	8.60, *26.41*	8.58, *26.30*	8.56, *26.23*	8.54, *26.18*	8.53, *26.12*
5.91, *14.37*	5.87, *14.24*		5.80, *14.02*	5.74, *13.83*	5.71, *13.74*	5.70, *13.69*	5.66, *13.57*	5.65, *13.52*	5.63, *13.46*
4.68, *9.89*	4.64, *9.77*		4.56, *9.55*	4.50, *9.38*	4.46, *9.29*	4.44, *9.24*	4.40, *9.13*	4.38, *9.07*	4.36, *9.02*
2.91, *4.71*	2.86, *4.60*		2.77, *4.41*	2.70, *4.25*	2.67, *4.17*	2.64, *4.12*	2.59, *4.01*	2.56, *3.96*	2.54, *3.91*
2.69, *4.16*	2.64, *4.05*		2.54, *3.86*	2.46, *3.70*	2.42, *3.61*	2.40, *3.56*	2.35, *3.46*	2.32, *3.41*	2.30, *3.36*
2.48, *3.67*	2.43, *3.56*		2.33, *3.36*	2.25, *3.20*	2.21, *3.12*	2.18, *3.07*	2.12, *2.97*	2.10, *2.92*	2.07, *2.87*
2.28, *3.23*	2.23, *3.13*		2.12, *2.94*	2.04, *2.77*	1.99, *2.69*	1.96, *2.63*	1.90, *2.53*	1.87, *2.47*	1.84, *2.42*
1.95, *2.56*	1.90, *2.46*		1.78, *2.26*	1.69, *2.10*	1.63, *2.00*	1.60, *1.94*	1.52, *1.82*	1.48, *1.76*	1.44, *1.68*
1.85, *2.36*	1.79, *2.26*		1.68, *2.06*	1.57, *1.89*	1.51, *1.79*	1.48, *1.73*	1.39, *1.59*	1.34, *1.51*	1.28, *1.43*
1.80, *2.28*	1.74, *1.17*		1.62, *1.97*	1.52, *1.79*	1.45, *1.69*	1.42, *1.62*	1.32, *1.48*	1.26, *1.39*	1.19, *1.28*
1.75, *2.18*	1.69, *2.07*		1.57, *1.87*	1.46, *1.69*	1.40, *1.59*	1.35, *1.52*	1.24, *1.36*	1.17, *1.25*	1.00, *1.00*

As with the t test, the F test is used to test the following hypotheses:

1. null hypothesis: H_0: $\sigma_1^2 = \sigma_2^2$ and
2. alternative hypothesis: H_1: $\sigma_1^2 > \sigma_2^2$,

where σ_1^2 and σ_2^2 are the variances of the two populations.

To reject or accept the null hypothesis, we compute the following F statistic:

$$F = \frac{s_1^2}{s_2^2},$$

where s_1^2 and s_2^2 are the sample variance estimates of the two populations. The ratio of two variances from two normal populations is also a random variable that follows an F distribution. The F distribution is illustrated in Figure 5.7. As with the t distribution, the F distribution varies with two parameters, such as the samples sizes of the two populations. Table 5.1 shows a fraction of an F table, where two degrees of freedom, dn and dd, are required to locate an F value in the F table. In this figure, the table entries are given for significance levels (α values) of 0.05 and 0.01. The F values associated with 0.05 and 0.01 significance levels are given in regular type and boldface-italics, respectively. Thus, for any two degrees of freedom, there are two F values provided, one for the 95% confidence level and one for the 99% confidence level.

To make use of the F table with the F test, we estimate an F statistic using the sample variance estimates from each of the two populations we are trying to compare. Note that for the use of this F table, the larger of the two variances should be put in the numerator of the equation above.

We now compare our estimated F statistic to the entries in the F table associated with dn and dd degrees of freedom and appropriate confidence level (only the 95% and 99% F values are provided in the table provided). $dn = n_1$ and $dd = n_2$ are the number of samples in each population, with n_1 being the number of samples in the population with variance placed in the numerator.

If we are to reject the null hypothesis outlined above, our calculated F statistic must be $> F(\alpha, dn, dd)$ in table to reject H_0 with confidence $(1 - \alpha) \times 100\%$. The degrees of freedom, $dn = n_1$, is the value used to locate the table entry in the horizontal direction (numerator), and $dd = n_2$ is the degrees of freedom used to locate the table entry in the vertical direction (denominator).

EXAMPLE 5.2 F test

	POPULATION A ($N_1 = 9$)	POPULATION B ($N_2 = 9$)
Mean	0.026	0.027
Variance	2.0E – 5	7.4E – 5

Given the data listed in table above, where we have collected 9 samples from each of two populations, A and B, we can estimate our F statistic to be

$$F = (7.4E - 5)/(2.0E - 5) = 3.7.$$

Using an F table that has table entries for both the 95% and 99% confidence levels, we find that the table entry for α equal to 0.05 and 9 df for both populations is

$$F(0.05, 9, 9) = 3.18.$$

Our estimated value of 3.7 exceeds the table entry of 3.18; thus, we may reject H_0 with 95% confidence and accept that population B has significantly greater variance than population A. However, we note that the table entry for $F(0.01, 9, 9)$ equals 5.35 and is greater than our estimated value. Thus, we cannot reject the null hypothesis with 99% confidence.

5.3 COMPARISON OF THREE OR MORE POPULATION MEANS

Thus far, we have discussed statistical analysis for comparing the means and variances (or standard deviations) for two populations based on collecting two sets of data. These two sets of data may be independent of each other or they may be dependent on each other, which we refer to as paired or blocked data. Another way to think of this pairing is to call it blocking on the experimental unit, meaning that both sets of data were collected from the same experimental units but under different conditions. For example, the experimental units may be human subjects, animals, medical instruments, manufacturing lines, cell cultures, electronic circuits, and more. If, for example, the experimental unit is human beings, the two sets of data may be derived before and after a drug has been administered or before and after the subjects have engaged in exercise. In another example, we may measure two different types of data, such as heart rate and blood pressure, each from the same group of subjects. This is referred to as repeated measures because we are taking multiple sets of measures from the same experimental units. In repeated measures, something in the experimental conditions, experimental unit, or measure being collected has changed from one set of measures to the next, but the sampling is blocked by experimental unit.

In many biomedical engineering applications, we need to compare the means from three or more populations, processes, or conditions. In such cases, we use a method of statistical analysis called analysis of variance, or ANOVA. Although the name implies that one is analyzing variances, the conclusions that stem from such analysis are in regard to significant difference in the means of three or more populations or processes.

5.3.1   One-Factor Experiments

We begin our discussions of ANOVA by discussing the design and analysis of one-factor experiments. In such experiments, we are drawing samples from three or more populations for which one factor has been varied from one population to the next.

As stated in Chapter 2, experimental design should include randomization and blocking. Randomization ensures that we do not introduce bias into the data because of ordering effects in collecting our data. Moreover, blocking helps us to reduce the effect of intersubject variability on the differences between populations.

Some biomedical engineering challenges that may require the use of ANOVA include the following:

1. Compare wear time between hip implants of various materials: In this example, the treatment = implant material (e.g., titanium, steel, polymer resins).
2. Compare MR pulse sequences in the ability to image tissue damage after stroke: In this example, the treatment = MR pulse sequence.
3. Compare ability of various drugs to reduce high blood pressure: In this example, the treatment = drug.

5.3.1.1   Example of Biomedical Engineering Challenge

An example of a biomedical engineering problem that might use ANOVA to test a hypothesis is in the study of reflex mechanisms in spinal cord injured patients [14]. One of the interests of investigators is how hip flexion (torque) is dependent on ankle movement in patients with spinal chord injury. In this example, there are actually two experimental factors that may affect hip flexion:

1. Range of ankle motion: in this case, the treatment = range of ankle extension.
2. Speed of ankle flexion: in this case, the treatment = speed of ankle extension.

We might evaluate one factor at a time using a one-factor or one-way ANOVA. Or, we might evaluate the impact of both factors on mean hip flexion using two-factor or two-way ANOVA. In either case, we have a null hypothesis and alternative hypotheses for the effect of each factor, such as range of ankle motion, on the population mean, such as hip flexion. Our null hypothesis is that there is no significant difference in mean hip flexion across ankle motion,

$$H_0: \mu_1 = \mu_2 = \mu_3 = \ldots = \mu_n.$$

The alternative hypothesis, H_1, states that at least two of the population means, such as hip flexion for two different ankle motions, differ significantly from each other.

TABLE 5.2: One-factor experiment

TREATMENT	OBSERVATIONS	SAMPLE MEAN	SAMPLE VARIANCE
1	$y_{11}, y_{22}, \ldots, y_{1n}$	$\bar{y}_1$	$\sigma^2 y_1$
2	$y_{21}, y_{22}, \ldots, y_{2m}$	$\bar{y}_2$	$\sigma^2 y_2$
$\vdots$			
i	$y_{i1}, y_{i2}, \ldots, y_p$	$\bar{y}_k$	$\sigma^2 y_i$
$\vdots$			
k	$y_{k1}, y_{k2}, \ldots y_{kq}$	$\bar{y}_k$	$\sigma^2 y_k$

The basic design of the one-factor experiment is given in Table 5.2. In this case, there are k treatments in one factor. k is the number of treatments, y_{ij} is the individual j sample or data point for the ith treatment, $\bar{y}_i$ is the sample mean for the ith treatment, and σ^2_{yi} is the sample variance for the ith treatment (adapted from [3]).

For example, we might be interested in comparing weight loss for three different types of diet pills. In this case, $k = 3$. Under the weight-loss column, one will find the weight loss of individual subjects who used one of the three diet pills. Note that in this example, the number of samples varies with each drug type. In a balanced experiment, there should be an equal number of samples for each type of treatment. In the third and fourth columns, we have the sample mean and standard deviation for weight loss for each of the diet pills. The question we are trying to answer is whether there is a significant difference in mean weight loss as a function of diet pill type.

DIET PILL	WEIGHT LOSS	MEAN	STANDARD DEVIATION
Placebo	10, 12, 5, 8, 5, 20	10.0	5.62
Drug B	30, 5, 12, 20	16.75	10.75
Drug C	2, 10, 5, 5, 10, 20, 25, 40	14.63	12.92

We make a number of assumptions when using ANOVA to compare differences in means between three or more populations or processes:

1. Subjects are randomly assigned to a specific treatment (in this case diet pill).
2. The populations or processes (such as weight loss) are approximately normally distributed.
3. Variance is approximately equal across treatment groups.

In this specific weight-loss example, no blocking is used. In other words, in no case did the same subject receive more than one treatment or diet pill. This may end up being problematic and lead to erroneous conclusions because we are not accounting for intersubject variability in response to the diet drug. For example, we might assume that the amount of weight loss is related to starting weight. It is possible that the subjects assigned to drug A all had starting weights less than the subjects given drug B. Thus, differences in the weight loss may be significant but may have little to do with the actual diet pill. The difference in weight loss may be because of the differences in starting weight rather than the diet pill. Because the above experiment does not use blocking, intersubject variability that is not accounted for may confound our conclusions regarding the effectiveness of diet pill.

The question we are attempting to address in using ANOVA is, "Are the means of the k treatments equal, or have our samples been drawn from treatments (populations) with different means?"

Within each treatment, i (or population, i), we assume that the variability across observed samples, Y_{ij}, is influenced by the population (treatment) mean, μ_i, and independent random variables, e_{ij}, that are normally distributed, with a mean of zero and variance σ^2. In other words, the samples, Y_{ij}, collected for each trial, j, within each treatment, i, can be expressed as

$$Y_{ij} = \mu_i + e_{ij}.$$

With our model, we are trying to determine how much of the variability in Y is because of the factor or treatment (population with mean, μ_i), and how much is because of random effects, e_{ij}, which we cannot control for or have not captured in the model provided above.

When we perform an ANOVA for a one-factor experiment, we can organize the analysis and results in the following table (Table 5.3):

TABLE 5.3: One-factor ANOVA with k treatments (no blocking)

SOURCE	df	SS	MS	F
Treatment	$k-1$			MS_{treat}/MS_{error}
Error	$N-k$			

N = total number of samples across all treatments; k = number of treatments within the one factor; F = the statistic that we will compare with our F tables (F distribution) to either reject or accept the null hypothesis; $SS_{treatment}$ = between-treatment sum of squares, a measure of the variability among treatment means; SS_{error} = within-treatment sum of squares, a measure of the sum of variances across all k treatments; $MS_{treatment}$ = $SS_{treatment}/k-1$; MS_{error} = $SS_{error}/N-k$.

For k treatments, the specific equations for the SS elements are the following:

$$SS_{treatment} = \sum_{i=1}^{k} n_i (\bar{y}_i - \bar{y}_{grand})^2$$

and

$$SS_{error} = \sum_{i=1}^{k} \sum_{j=1}^{n_i} (Y_{ij} - \bar{Y}_i)^2 = \sum_{i=1}^{k} \sigma_{yi}^2.$$

$\bar{Y}_{grand}$ is the sample mean for all samples across all treatments combined.

Finally, the F statistic that we are most interested in estimating to reject or accept our null hypothesis is

$$F = MS_{treatment}/MS_{error}.$$

To test our hypothesis at a significance level of α, we now compare our estimated F statistic to the F distribution provided in the F tables for the table entry, $F(\alpha; k-1, N-k)$. If our estimated F value is greater than the table entry, we may reject the null hypothesis with $(1-\alpha) \times 100\%$ confidence.

Example 5.3 A heart valve manufacturer has three different processes for producing a leaf valve. Random samples of 50 valves were selected five times from each type of manufacturing process. Each valve was tested for defective opening mechanisms. The number of defective valves in each sample of 50 valves is summarized in the following table:

PROCESS A	PROCESS B	PROCESS C
1	5	3
4	8	1
3	6	1
7	9	4
5	10	0

Using an $\alpha = 0.05$, we want to determine whether the mean number of defects differs between processes (treatment). (Our null hypothesis, H_0, is that mean number of defects is the same across processes.)

To answer the question, we need to complete the following ANOVA table.

One-way ANOVA: processes A, B, and C				
SOURCE	df	SS	MS	f
Factor	2			
Error	12			
Total	14			

We first find SS_{factor} (= $SS_{treatment}$).
 We know that

$$SS_{treatment} = \sum_{i=1}^{k} n_i (\bar{y}_i - \bar{y}_{grand})^2.$$

From the data given, we know that there are three treatments, A, B, and C; thus, $k = 3$. We will assign $A = 1$, $B = 2$, and $C = 3$. We also know that $n_1 = n_2 = n_3 = 5$. We can estimate the sample mean for each treatment or process to obtain, $\bar{y}_1 = 4$, $\bar{y}_2 = 7.6$, $\bar{y}_3 = 1.8$. We can also use all 15 samples to find $\bar{y}_{grand} = 4.47$.

Now, we can use these estimates in the equation for $SS_{treatment}$:

$$SS_{treatment} = 5(4 - 4.47)^2 + 5(7.6 - 4.47)^2 + 5(1.8 - 4.47)^2 = 85.70.$$

Now we solve for SS_{error}:
 Given that

$$SS_{error} = \sum_{i=1}^{k} \sum_{j=1}^{n_i} (Y_{ij} - \bar{Y}_i)^2,$$

we need to estimate the inner summation for each treatment, i, noted by the outer summation. Thus, for $i = 1$,

$$SS_1 = (1 - 4)^2 + (4 - 4)^2 + (3 - 4)^2 + (7 - 4)^2 + (5 - 4)^2 = 20;$$

$$\text{for } i = 2, \ SS_2 = (5 - 7.6)^2 + (8 - 7.6)^2 + (6 - 7.6)^2 + (9 - 7.6)^2 + (10 - 7.6)^2 = 17.2;$$

$$\text{for } i = 3, \ SS_3 = (3 - 1.8)^2 + (1 - 1.8)^2 + (1 - 1.8)^2 + (4 - 1.8)^2 + (0 - 1.8)^2 = 10.8.$$

Now, summing over all k treatments, where $k = 3$, we find

$$SS_{error} = SS_1 + SS_2 + SS_3 = 48.0.$$

Now, we can find the mean squared errors:

$$MS_{treatment} = SS_{treatment}/DF_{treatment} = 87.5/2 = 43.7.$$

$$MS_{error} = SS_{error}/DF_{error} = 48.0/12.0 = 4.$$

Finally, our F statistic is given by

$$F = MS_{treatment}/MS_{error} = 43.7/4 = 10.92.$$

We now complete our ANOVA table.

One-way ANOVA: processes A, B, and C				
SOURCE	*df*	**SS**	**MS**	*f*
Factor	2	87.5	43.7	10.92
Error	12	48.0	4.0	

Now, we want to determine if the mean number of defects differ among processes A, B, and C. Our null hypothesis is that the mean number of defects does not differ among processes. To reject this hypothesis at the 95% confidence level ($\alpha = 0.05$), our estimated F statistic must be greater than $F(0.05, 2, 12) = 3.88$, found in the F distribution tables. Our F statistic = 10.92 > 3.88; thus, we can reject the null hypothesis with 95% confidence and conclude that the number of defects does vary with manufacturing process and that our samples for treatments A, B, and C were drawn from populations with different means. In fact, the F distribution table list a value for $F(0.01, 2, 12) = 6.93$. Thus, we can reject the null hypothesis at the 99% confidence level as well.

Example 5.4 Four types of MR scanners are being evaluated for speed of image acquisition. The following table summarizes the speeds measured (in minutes) for three samples of each scanner type.

SCANNER A	SCANNER B	SCANNER C	SCANNER D
2.0	4.0	3.0	6.0
1.8	4.5	2.5	5.5
2.7	5.5	2.0	3.5

Does the mean speed of image acquisition differ amongst the four scanner types? (Use $\alpha = 0.01$.)

To answer this question, we may work through the calculations that we performed in Example 5.3. However, as one discovered in the previous example, as the sample size grows, the number of calculations quickly grows. Thus, most investigators will use a statistical software package to perform the ANOVA calculations. For this example, we used the Minitab software package (Minitab Statistical Software, Release 13.32, Minitab, 2000) to perform the ANOVA. ANOVA produces the following table:

ANOVA for speed					
SOURCE	*df*	SS	MS	*f*	α
Factor	3	19.083	6.361	9.07	0.006
Error	8	5.613	0.702		
Total	11	24.697			

The F statistic of 9.07 results in an area with $\alpha = 0.006$ in the right tail of the F distribution. Because $\alpha < 0.01$, the value of α at which we were testing our hypothesis, we can reject the null hypothesis and accept that alternative hypothesis that at least one of the scanners differs from the remaining scanners in the mean speed of acquisition.

In the examples given previously, we did not use blocking in the experimental design. In other words, the populations from which we collected samples differed from treatment to treatment. In some experiments, such as testing weight loss for diet pills, it is not practical or possible to test more than one type of treatment on the same experimental unit.

However, when blocking may be used, it should be used to compare treatments to reduce intersubject variability or differences (that cannot be controlled for) in the experimental outcome. Table 5.4 outlines the experimental design for a one-factor experiment that makes use of blocking. In this experimental design, all experimental units are subject to the every treatment. The result is

TABLE 5.4: One factor with k treatment and blocking

TREATMENT (ANKLE SPEED)	SUBJECT NUMBER				TREATMENT MEAN
	1	2	3	4	
10	5	2	5	6	
20	10	2	8	4	
30	15	8	15	10	
Block mean					

that we now have treatment means and block means. Thus, when we perform the ANOVA analysis, we may test two different sets of hypotheses. The first null hypothesis is that there is not significance difference in treatment means. The second null hypothesis is that there is no significant difference in means across subjects or experimental units (intersubject variability is not significant).

In this example, treatment could be ankle speed, and the block would be the patient, such that each ankle speed is tested on each and every patient subject. In such a design, it is important to randomize the order in which different ankle speeds are tested so that there is no bias in hip flexion due to ordering effects of ankle speed or machine wear.

Note that the example above is also a balanced experimental design because there are same numbers of data points in every cell of the table.

Within each treatment, i (or population, i), we now assume that the variability across observed samples, Y_{ij}, is influenced by the population (treatment) mean, μ_i, the block effects, β_j, and independent random variables, e_{ij}, which are normally distributed with a mean of zero and variance σ^2. In other words, the samples, Y_{ij}, collected for each trial, j, within each treatment, i, can be expressed as

$$Y_{ij} = \mu_i + \beta_j + e_{ij}.$$

In other words, we are trying to determine how much of the variability in Y is because of the factor or treatment (population with mean, μ_i) and how much is due to block effects, β_j, and random effects, e_{ij}, that we cannot control for or have not captured in the model provided above. In such a model, we assume that treatment effects and block effects are additive. This is not a good assumption when there are interaction effects between treatment and block. Interaction effects mean

TABLE 5.5: One-factor ANOVA with block

SOURCE	df	SS	MS	F
Treatment	$k - 1$			MS_{treat}/MS_{error}
Block	$b - 1$			MS_{block}/MS_{error}
Error	$(b - 1)(k - 1)$			

that the effect of a specific treatment may depend on a specific block. When we address two-factor experiments in the next section, we will discuss further these interaction effects.

We can summarize the ANOVA for one factor with block with the following ANOVA table (assuming no interaction effects) (Table 5.5; adapted from [3]).

Note that we now have two estimated F values to consider. $F_{treatment}$ is compared with the table entry for $F(\alpha; k - 1, (b - 1)(k - 1))$, and the null hypothesis stating that there is no difference in mean hip flexion across treatments is rejected if the estimated F value is greater than the table entry.

F_{block} is compared with the table entry for $F(\alpha; b - 1, (b - 1)(k - 1))$, and the null hypothesis stating that there is no difference in mean hip flexion across subjects is rejected if the estimated F statistic is greater than the table entry.

Example 5.5 One-factor experiment with block

HeartSync manufactures four types of defibrillators that differ in the strength of the electrical shock given for an episode of fibrillation. A total of 280,000 patients were divided into four groups of 70,000 patients each. Each group was assigned to one of the four defibrillators, and the number of shocks that failed to defibrillate was recorded for four consecutive years. The results were as follows:

YEAR AFTER IMPLANT	DEVICE A	DEVICE B	DEVICE C	DEVICE D
1	6	1	9	2
2	8	1	10	2
3	5	3	8	0
4	10	2	11	5

There are two questions we wish to address with these data:

1. Using $\alpha = 0.01$, does the mean number of failures differ significantly as a function of device type?
2. Using $\alpha = 0.01$, does the mean number of failures differ significantly as a function of year after implant? In other words, is year after implant a major source of variability between populations?

Again, rather than estimate the calculations by hand, we may use statistical software, such as Minitab, to obtain the following results using ANOVA with block:

ANOVA for number of failures				
SOURCE	*df*	SS	MS	*F*
Device	3	173.188	57.729	35.67
Year	3	20.688	6.896	4.26
Error	9	14.563	1.618	
Total	15	208.438		

The F statistic for device, $F = 35.67$, is greater than the critical F value, $F(0.01, 3, 9) = 6.99$, given in the F table. Thus, we can reject the null hypothesis and accept that alternative hypothesis that at least one of the devices differs from the remaining devices in the mean number of failure. The second part of the questions tests the hypothesis that the mean number of failures differs significantly as a function of year after implant (the blocking factor). The F statistic for year, $F = 4.26$, is less than the critical F value, $F(0.01, 3, 9) = 6.99$, given in the F table; thus, we accept our null hypothesis, which means failure rate does not differ between years after implant.

5.3.2   Two-Factor Experiments

In the original biomedical engineering challenge describing hip flexion reflexes, we discussed two factors that influence hip flexion: ankle speed and range of ankle extension.

In a two-factor experiment leading to a two-way ANOVA, there are two factors being varied, A and B, where A has a treatments and B has b treatments, and there are n samples at every combination of A and B.

The two-factor experiment is said to be completely crossed if there are samples collected for every combination of factors A and B. In addition, the experiment is said to be balanced if we have same number of samples for every combination of factors A and B.

Table 5.6 below illustrates a two-factor experiment that is completely balanced and crossed. Each of the two treatments within factor A is crossed with each of the three treatments within factor B. Also note that for each combination of A and B, we have three samples.

The question that we are trying to address with a two-factor ANOVA is whether there are differences in treatment means for each of the two factors. In addition, we wish to know if there are interaction effects such that there are significant differences in means as a function of the cross-interaction between factors. In other words, there are significant differences in sample means, and hence, the means of the underlying populations, when specific combinations of factors A and B occur together.

Once the data have been collected, as illustrated in Table 5.6 above, we may perform a two-factor ANOVA to test the following three null hypotheses:

$$H_0: \mu_{A1} = \mu_{A2} = \mu_{A3} = \dots. = \mu_{Aa};$$

$$H_0: \mu_{B1} = \mu_{B2} = \mu_{B3} = \dots. = \mu_{Bb};$$

$$H_0: \mu_{A1B1} = \mu_{A1B2} = \mu_{A1B3} = \mu_{A2B1} = \mu_{A2B2} = \dots = \mu_{AaBb}.$$

For each of the three null hypotheses, the associated alternative hypothesis is that there is a significant difference in at least two of the population means for a given factor or combination of factors.

The analysis and results of a two-factor ANOVA may be organized as in Table 5.7 [3].

The equations for the SS and MS for each of the factors and interaction factors are beyond the scope of this text but may be found in [3]. In practice, the investigator will use a popular statistical software package such as Minitab, SPSS, or SAS to estimate these SS and MS values (because of computational burden) and simply refer to the F statistics to reject or accept the null hypotheses.

TABLE 5.6: Two-factor experiment

	FACTOR B		
Factor A	1	2	3
1	1.2, 1.4, 2.1	2.3, 2.2, 2.6	6.4, 5.8, 3.2
2	3.2, 4.1, 3.6	4.1, 4.3, 4.0	8.2, 7.8, 8.3

TABLE 5.7: Two-factor ANOVA table (each with multiple treatments)

SOURCE	df	SS	MS	F
A	$a-1$			MS_A/MS_{error}
B	$b-1$			MS_B/MS_{error}
AB	$(a-1)(b-1)$			MS_{AB}/MS_{error}
Error	$ab(n-1)$			

We note that there are three F statistics to test each of the three null hypotheses described earlier. In each case, we compare our estimated F statistic with the F values in our F table representing the F distribution. More specifically, we compare our estimates F values to the following table entries:

To test H_0: $\mu_{A1} = \mu_{A2} = \mu_{A3} = ... = \mu_{Aa}$, compare F_A with $F(\alpha; a-1, ab(n-1))$;

To test H_0: $\mu_{B1} = \mu_{B2} = \mu_{B3} = ... = \mu_{Bb}$, compare F_B with $F(\alpha; b-1, ab(n-1))$;

To test H_0: $\mu_{A1B1} = \mu_{A1B2} = \mu_{A1B3} = \mu_{A2B1} = \mu_{A2B2} = ... = \mu_{AaBb}$,

compare F_{AB} with $F(\alpha; (a-1)(b-1), ab(n-1))$.

In each of the three tests, if the estimated F statistic is greater than the table entries for the F distribution, one may reject the null hypothesis and accept the alternative hypothesis with $(1-\alpha) \times 100\%$ confidence.

Example 5.6 An example of a two-factor experiment that will be evaluated using two-factor ANOVA occurs when biomedical engineers are looking at the effectiveness of rehabilitative therapy and pharmacological therapy on the recovery of movement in a limb after stroke. Details of the experimental design include the following:

1. Factor T: therapy used (there are three types of therapies, T1, T2 and T3);
2. Factor D: drug used (there are three types of Drugs, D1, D2 and D3);
3. 36 patients are randomly assigned to each combination of T and D;
4. measure: number of days to meet recovery criteria.

Experimental design for two factors								
T1			**T2**			**T3**		
D1	D2	D3	D1	D2	D3	D1	D2	D3
20	25	13	22	8	16	9	15	7
15	16	12	16	10	19	12	10	10
18	10	22	17	9	11	8	9	9
24	20	10	12	11	21	8	10	9

The questions we are trying to address is whether there is a significant difference in mean days of recovery for the three types of rehabilitative therapy, the mean days of recovery for the three types of drug therapy, and differences in mean days of recovery for the nine combinations of rehabilitative therapy and drug therapy (interaction effects).

ANOVA analysis performed using statistical software, known as Minitab, produces the following table summarizing the ANOVA analysis. Note that there are three estimated F statistics. We can use the three F values to test our hypotheses:

$$H_0: \mu_{T1} = \mu_{T2} = \mu_{T3};$$

$$H_0: \mu_{D1} = \mu_{D2} = \mu_{D3};$$

$$H_0: \mu_{T1D1} = \mu_{T1D2} = \mu_{T1D3} = \mu_{T2B1} = \mu_{T2B2} = \cdots = \mu_{T3D3}.$$

Two-way ANOVA for days of recovery versus T and D					
SOURCE	*df*	**SS**	**MS**	*F*	α
T	2	337.4	168.7	11.44	0.000
D	2	36.2	18.1	1.23	0.309
Interaction	4	167.8	41.9	2.84	0.043
Error	27	398.3	14.8		

Note that the estimated F statistic for rehabilitative therapy, $F = 11.44$, does exceed the critical value for $\alpha < 0.05$ (actually, α is <0.0001); thus, we conclude that the samples for the three different rehabilitative therapies represent populations with different means. In other words, the different rehabilitative therapies produce different days of recovery. However, the estimated F statistic for the drug therapy does not exceed the critical table F value for $\alpha = 0.05$; thus, we accept the null hypothesis that drug therapy does not significantly affect the days of recovery and that the samples are not drawn from populations with different means. Moreover, the third F statistic, $F = 2.84$, is greater than the table F value for $\alpha < 0.05$, which suggests that there may be a significant difference in days of recovery due to an interaction between rehabilitative therapy and drug therapy.

5.3.3   Tukey's Multiple Comparison Procedure

Once we have established that there is a significant difference in means across treatments within a factor, we may use post hoc tests, such as the Tukey's HSD multicomparison pairwise test [3, 9]. ANOVA simply shows that there is at least one treatment mean that differs from the others. However, ANOVA does not provide information on specifically which treatment mean(s) differs from which treatment mean(s). The Tukey's HSD test allows us to compare the statistical differences in means between all pairs of treatments. For a one-factor experiment with k treatments, there are $k(k - 1)/2$ pairwise comparisons to test. The important point to note is that if α is the probability of a type I error for one comparison, the probability of making at least a type I error for multiple comparisons is much greater. So, if we want $(1 - \alpha) \times 100\%$ confidence for all possible pairwise comparisons, we must start with a much smaller α. Tukey's multiple comparison procedure allows for such an adjustment in significance when performing pairwise comparisons.

• • • •

CHAPTER 6

Linear Regression and Correlation Analysis

Oftentimes, in biomedical engineering research or design, we are interested in whether there is a correlation between two variables, populations, or processes. These correlations may give us information about the underlying biological processes in normal and pathologic states and ultimately help us to model the processes, allowing us to predict the behavior of one process given the state of another correlated process.

Given two sets of samples, X and Y, we ask the question, "Are two variables or random processes, X and Y, correlated?" In other words, can y be modeled as a linear function of x such that

$$\hat{y} = mx + b.$$

Look at the following graph of experimental data (Figure 6.1), where we have plotted the data set, y_i, against the data set, x_i:

We note that the data tend to fall on a straight line. There tends to be a trend such that y increases in proportion to increases in x. Our goal is to determine the line (the linear model) that best fits these data and how close the measured data points lie with respect to the fitted line (generated by the model). In other words, if the modeled line is a good fit to the data, we are demonstrating that y may be accurately modeled as a linear function of x, and thus, we may predict y given x using the linear model.

The key to fitting a line that best predicts process y from process x, is to find the parameters m and b, which minimize the error between model and actual data in a least-squares sense:

$$\min [(y - \hat{y})^2].$$

In other words, as illustrated in Figure 6.2, for each measured value of the independent variable, x, there will be the measured value of the dependent variable, y, as well as the predicted or

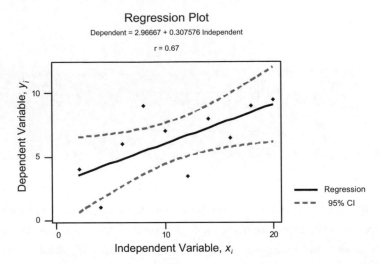

FIGURE 6.1: Results of linear regression applied to the samples illustrated in the scatterplot (black dots). The solid black line illustrates the line of best fit (model parameters listed above the graph) as determined by linear regression. The red dotted curves illustrate the confidence interval for the slope. Finally, the r value is the correlation coefficient.

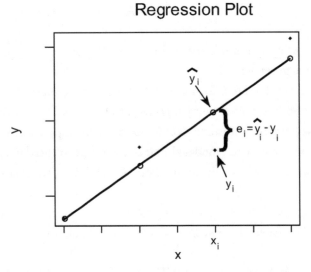

FIGURE 6.2: Linear regression can be used to estimate a straight line that best "fits" the measured data points (filled circles). In this illustration, x_i and y_i represent the measured independent and dependent variables, respectively. Linear regression is used to model the dependent variable, y, as a linear function of the independent variable, x. The straight line passing through the measured data points is the result of linear regression whereby the error, e_i, between the predicted value (open circles) of the dependent variable, $\hat{y}_i$, and the measured value of the dependent variable, y_i, is minimized over all data points.

modeled value of y, denoted as $\hat{y}$, that one would obtain if the equation $\hat{y} = mx + b$ is used to predict y. The equation $e_i = y_i - \hat{y}_i$ denotes the errors that occur at each pair of (x_i, y_i) when the modeled value does not exactly align with the predicted value because of factors not accounted for in the model (noise, random effects, and nonlinearities).

In trying to fit a line to the experimental data, our goal is to minimize these e_i between the measured and predicted values of the dependent variable, y. The method used in linear regression and many other biomedical modeling techniques is to find model parameters, such as m and b, that minimize the sum of squared errors, e_i^2.

For linear regression, we seek a least-squares estimate of m and using the following approach:

Suppose we have N samples each of processes x and y. We try to predict y from measured x using the following model:

$$\hat{y} = mx + b.$$

The error in prediction at each data point, x_i, is

$$\text{error}_i = y_i - \hat{y}_i.$$

In the least-squares method, we choose m and b to minimize the sum of squared errors:

$$(y_i - \hat{y}_i)^2 \qquad\qquad\qquad \text{for } i = 1 \text{ to } N.$$

To find a closed form solution for m and b, we can write an expression for the sum of squared errors:

$$\sum_{i=1}^{N} e_i^2 = \sum_{i=1}^{N} (y_i - \hat{y}_i)^2.$$

We then replace $\hat{y}_i$ with $(mx_i + b)$, our model, and carry out the squaring operations [3, 5].

We can then take derivatives of the above expression with respect to m and then again with respect to b. If we set the derivative expressions to zero to find the minimums, we will have two equations in two unknowns, m and b, and we can simply use algebra to solve for the unknown parameters, m and b. We will get the following expressions for m and b in terms of the measured x_i and y_i:

$$m = \frac{\displaystyle\sum_{i=0}^{N-1} x_i y_i - \left(\sum_{i=0}^{N-1} x_i\right)\left(\sum_{i=0}^{N-1} y_i\right) / N}{\displaystyle\sum_{i=0}^{N-1} x_i^2 - \left(\sum_{i=0}^{N-1} x_i\right)^2 / N}$$

and

$$b = \bar{y} - m\bar{x},$$

where

$$\bar{x} = \frac{1}{N}\sum_{i=0}^{N-1} x_i \quad \text{and} \quad \bar{y} = \frac{1}{N}\sum_{i=0}^{N-1} y_i.$$

Hence, once we have our measured data, we can simply use our equations for m and b to find the line, or linear model, of best fit.

The Correlation Coefficient

It is important to realize that linear regression will fit a line to any two sets of data regardless of how well the data are modeled by a linear model. Even if the data, when plotted as a scatterplot, look nothing like a line, linear regression will fit a line to the data. As biomedical engineers, we have to ask, "How well does the measured data "fit" the line estimated through linear regression?"

One measure of how well the experimental data fit the linear model is the correlation coefficient. The correlation coefficient, r, has a value between –1 and 1 and indicates how well the linear model fits to the data.

The correlation coefficient, r, may be estimated from the experimental data, x_i and y_i, using the following equation:

$$r = \frac{\sum\limits_{i=0}^{N-1}\left(x_i - \bar{x}\right)\left(y_i - \bar{y}\right)}{\left[\sum\limits_{i=0}^{N-1}\left(x_i - \bar{x}\right)^2 \sum\limits_{i=0}^{N-1}\left(y_i - \bar{y}\right)^2\right]^{1/2}},$$

where

$$\bar{x} = \frac{1}{N}\sum_{i=0}^{N-1} x_i \quad \text{and} \quad \bar{y} = \frac{1}{N}\sum_{i=0}^{N-1} y_i.$$

It is important to note that an $r = 0$ does not mean that the two processes, x and y, are independent. It simply indicates that any dependency between x and y is not well described or modeled by a linear relation. There could be a nonlinear relation between x and y. An $r = 0$ simply means that x and y are uncorrelated in a linear sense. That is, one may not predict y from x using a linear model, $y = mx + b$.

A measure related to the correlation coefficient, r, is the coefficient of determination, R^2, which is a summary statistic that tells us how well our regression model fits our data. R^2 can be used as measure of goodness of fit for any regression model, not just linear regression. For linear regression, R^2 is the square of the correlation coefficient and has a value between 0 and 1. The coefficient of determination tells us how much of the variability in the data may be explained by the model parameters as a fraction of total variability in the data.

It is important to realize that the estimated slope of best fit and the correlation coefficient are statistics that may or may not be significant. Thus, t tests may be performed to test if the slope estimated through linear is significantly different from zero [3]. Likewise, t tests may be performed to test if the correlation coefficient is significantly different from zero. Finally, we may also compute confidence intervals for the estimated slope [3].

• • • •

CHAPTER 7

Power Analysis and Sample Size

Up to this point, we have discussed important aspects of experimental design, data summary, and statistical analysis that will allow us to test hypotheses and draw conclusions with some level of confidence.

However, we have not yet addressed a very important question. The question we ask now is, "how large should my sample be to capture the variability in my underlying population so that my types I and II error rates will be small?" In other words, how large of a sample is required such that the probability of making a type I or II error in rejecting or accepting a null hypothesis will be acceptable under the circumstances. Different situations call for different error rates. An error such as diagnosing streptococcal throat infection when streptococcus bacteria are not present is likely not as serious as missing the diagnosis of cancer. Another way of phrasing the question is, "How powerful is my statistical analysis in accepting or rejecting the null hypothesis?"

If the sample size is too small, the consequence may be that we miss an effect (type II error) because we do not have enough power in the test to demonstrate, with confidence, an effect.

However, when choosing a sample size, it is too easy to simply say that the sample size should be as large as possible. Even if an investigator had access to as many samples as he or she desired, there are practical considerations and constraints that limit the sample size. If the sample size is too large, there are economic and ethical problems to consider. First, there are expenses associated with running an experiment, such as a clinical trial. There are costs associated with the personnel who run the experiments, the experimental units (animals, cell cultures, compensation for human time), perhaps drugs and other medical procedures that are administered, and others. Thus, the greater the number of samples, the greater the expense. Clinical trials are typically very expensive to run.

The second consideration for limiting sample size is an ethical concern. Many biomedical-related experiments or trials involve human or animal subjects. These subjects may be exposed to experimental drugs or therapies that involve some risk, and in the case of animal studies, the animal may be sacrificed at the end of an experiment. Bottom line, we do not wish to use human or animal subjects for no good reason, especially if we gain nothing in terms of the power of our statistical analysis by increasing the sample size.

We recall again that there are two types of errors associated with performing a statistical analysis:

1. Type I: rejecting H_0 when H_0 is true (probability of type I error = α);
2. Type II: accepting H_0 when H_0 is false; missing an effect (probability of type II error = β).

7.1   POWER OF A TEST

When we refer to the power of a given statistical analysis or test, we are quantifying the likelihood, for a given α value (e.g., 0.05), that a statistical test (t test, correlation coefficient, ANOVA, etc.) will detect a real effect. For example, if there truly is a difference between the means of two populations, what is the likelihood that we will detect that difference with our t test? If the chance of a type II error is β, then the likelihood of detecting a true difference is simply $1 - \beta$, which we refer to as the power of the test. Some practical points about the power of a statistical test are the following:

1. power of test = $1 - \beta$;
2. $0 <$ power < 1
3. In general, we desire power ≥ 0.8 for practical applications.

When trying to establish a sample size for an experiment, we have to decide in advance how much power we want for our statistical analysis, which, in turn, is determined by the amount of types I and II errors we are willing to risk. For some biomedical engineering applications, we can risk greater error than others. Missing a diagnosis may not be problematic in one case or may result in a life-threatening situation. If an implantable defibrillator fails to detect and electrically shock a heart rhythm, such as ventricular fibrillation, it could mean the loss of life. On the other hand, shocking the heart when there is no fibrillation present, which may happen when the device senses electrical activity generated by nearby skeletal muscle, can result in pain and may even initiate a dangerous heart rhythm.

It is often difficult to minimize both α and β at the same time. One is usually minimized at the expense of another. For example, we would rather error on the side of an implantable defibrillator overdetecting fibrillation so as not to miss an occurrence of fibrillation. In the case of detecting breast cancer using mammography, overdetection or underdetection can both be problematic. Underdetection can mean increased chance of death. However, false-positives can lead to unnecessary removal of healthy tissue. One of the biggest challenges for researchers in the biomedical field is to try to find real differences between populations that will improve diagnosis and detection of disease or the performance of medical instruments.

7.2   POWER TESTS TO DETERMINE SAMPLE SIZE

Power tests are used to determine sample size and take into account the effect that we wish to detect with our statistical analysis, the types I and II error rates we will tolerate, and the variability of the populations being sampled in our experiments.

To perform a power test, we can use equations that express power in terms of the factors stated above, or we can make use of power curve and power tables that have already estimated the sample sizes for us. Power curves and tables show the relationship between power $(1 - \beta)$ and the effect we are trying to detect. These effects can be difference in two means, difference in two variances, a correlation coefficient, difference in treatment means, and other population differences that we are trying to detect through experiment and data analysis.

It is important to note that there are different equations and, thus, different power curves for different effects, which in turn are detected through different statistical tests. A difference in two means (the effect) is detected with a t test, whereas the difference in two variances (the effect) is detected using an F test. The power curves for the t test are different from the power curves for the F test.

Power curves most frequently used in estimating a sample size for a biomedical experiment in include the following:

1. unpaired or paired t test (difference in one or two population means);
2. Pearson's correlation coefficient (correlation between two populations);
3. ANOVA (difference in three or more population means).

To perform a power test using the power curves, we need the following:

1. the size of effect we want to detect (i.e. difference in two means);
2. an estimate of population parameters (i.e. standard deviation for the population(s) based on pilot data);
3. α level (probability of type I error);
4. power level $= (1 - \beta)$ (probability of type II error).

The α and β levels are selected by the investigator before designing the experiment. The size of effect to detect is also chosen by the investigator. The investigator needs to determine in advance how large the effect has to be to be significant for drawing conclusions. In other words, how different do two populations means need to be to signal something different in the underlying physiology, drug effects, or a change in the manufacturing process. Or, how strong does the correlation have to be to mean something to the investigator, in terms of underlying biology or processes? In general, the smaller the difference to be detected with a *t* test or the greater the correlation value to be detected, the greater the sample size.

One might argue that the most difficult piece of information to obtain is an estimate of the standard deviation, or variance, of the underlying populations. It is the assumed variance of the underlying population that plays a large role in determining sample size. In general, the larger the variance in the population, the greater the sample size needed to capture the statistics of the population or process and, consequently, the greater the sample size needed to give power to a statistical analysis.

Table 7.1 shows an example of a power table for the unpaired *t* test. This table was estimated for $\alpha = 0.05$. But similar tables exist for other α levels. These curves and tables allow us to estimate a sample size if we first select a normalized difference in the two means that we wish to detect and we select a power at which to perform the *t* test. The normalized difference is obtained when we take the absolute difference that we wish to detect and divide that difference by the estimated standard deviation of the underlying population. Again, we have a guess for the standard deviation based on pilot data. By normalizing the difference to detect, we need not worry about units of measure and can make used of standardized tables.

TABLE 7.1: Table for power for Student's unpaired (one-tail) *t* test ($\alpha = 0.05$)

	Difference in means (expressed as z score)									
n	0.10	0.20	0.30	0.40	0.50	0.60	0.70	0.80	1.0	1.2
10	0.08	0.11	0.16	0.22	0.29	0.36	0.45	0.53	0.70	0.83
13	0.08	0.13	0.18	0.26	0.34	0.44	0.54	0.63	0.80	0.91
20	0.09	0.15	0.24	0.34	0.46	0.59	0.70	0.80	0.93	0.98
30	0.10	0.19	0.31	0.46	0.61	0.74	0.85	0.92	0.99	
40	0.11	0.22	0.38	0.55	0.72	0.84	0.93	0.97		
60	0.13	0.29	0.50	0.70	0.86	0.95	0.98			
80	0.15	0.35	0.60	0.81	0.93	0.98				
100	0.17	0.41	0.68	0.88	0.97					
200	0.26	0.64	0.91	0.99						

The best way to illustrate the use of power tests is to work through an example. We will use the table in Figure 7.1 to find the sample size for the following example:

Aim: To determine if there is there a difference in mean height for college-age men and women living in the city of make-believe?

Experiment: Measure the heights in random samples of college-age men and women in the city of make-believe.

Question: How many samples do we need to reject our null hypothesis that there is no difference in mean height between college-age men and women in the city of make-believe if indeed there is a true difference in the two populations?

We need to perform a power test to determine how many samples we need to collect from the college-age men and women if we are to detect a true difference in the underlying populations. We are assuming that the two underlying populations are normally distributed. The effect that we are trying to detect is whether there is a significant difference in mean heights; thus, we will be using an unpaired t test to analyze the data once they are collected.

Before performing the power test, we need to decide on the magnitude of effect we wish to detect as being significant when we perform our t test. For this example, we are going to choose a 3-in. difference in mean heights as being significant. For argument sake, we will claim that a 3-in. difference in population means is of biological significance.

In this example, the effect being tested ($\mu_1 - \mu_2$) is 3 in. To use the standardized curves, we need to normalize this difference by an estimate of the standard deviation, σ, of the underlying populations. This is basically a z score for the difference in the two means. Note that we assume the two populations to have roughly equal variance if we are to use a t test. To obtain an estimate of σ, let us assume we have collected some pilot data and estimated the sample standard deviation, $s = 5$ in. Now, the normalized difference we wish to detect when we apply our t test is 3/5.

We next choose $\alpha = 0.05$ and power = 0.8. Again, these error rates are our choice.

Now, using the magnitude of our effect (3/5), power value (0.8), and α value (0.05), we used the tables or curves in Figure 7.1 to look up the number of required samples from the power tables. For this example, we find the sample size to be approximately 35 samples.

Some general observations may be made about the impact of population variance, types I and II error rates, and the magnitude of the effect on the sample size:

1. For a given power and variance, the smaller the effect to detect, the greater the sample size.
2. The greater the power, the greater the sample size.
3. The greater the population variance, the greater the sample size.

If we use the same example above and the Minitab software to calculate sample size for an unpaired t test as we vary several of the inputs to the power test, we find the following sample sizes (Table 7.2):

TABLE 7.2: Relations among sample size, statistical power, standard deviation, and magnitude of effect to detect ($\alpha = 0.05$)

Size of difference in means (in.)	Standard deviation (in)	Power of test	Size of sample
1	5	0.8	394
2	5	0.8	100
3	5	0.8	45
4	5	0.8	26
3	1	0.8	4
3	2	0.8	9
3	10	0.8	176
3	5	0.9	60

Thus, we can estimate, in advance of performing an experiment, a minimal sample size that would allow us to draw conclusions from our data with a certain level of confidence and power.

•   •   •   •

CHAPTER 8

Just the Beginning…

Compared with most textbooks about statistics, the previous seven chapters have provided a brief "elevator pitch" about the fundamentals of statistics and their use in biomedical applications. The reader should be familiar with the basic notion of probability models, the use of such models to describe real-world data, and the use of statistics to compare populations. The reader should be aware of the impact of randomization and blocking on the outcome of statistical analysis. In addition, the reader should have an appreciation for the importance of the normal distribution in describing populations, the use of standardized tables, and the notion that statistical analysis is aimed at testing a hypothesis with some level of confidence. Finally, the reader should know that we can find confidence intervals for any estimated statistic.

In this text, we only covered those statistical analyses that are valid for populations or processes that are well modeled by a normal distribution. Of course, there are many biological processes that are not well modeled by a normal distribution. For other types of distributions, one should read more advanced texts to learn more about nonparametric statistics and statistics for nonnormal distributions or populations. These nonparametric tests do not assume an underlying distribution for the data and are often useful for small size samples.

How do we determine whether our data, and hence underlying population, is well modeled by a normal distribution? We can begin by simply looking at the histogram of the sampled data for symmetry and the proportion of samples that lie within one, two, and three standard deviations of the mean. We can also compare the sample mean to the sample median. A more formal means of quantifying the normality of a sample is to use a χ^2 test [7]. The χ^2 test involves comparing the actual sample distribution to the distribution that would be expected if the sample were drawn from a normal distribution. The differences between the expected frequencies of sample occurrence and the true frequencies of sample occurrence are used to estimate a χ^2 test statistic. This test statistic is then compared with the critical values of the χ^2 distribution to determine the level of significance or confidence in rejecting the null hypothesis. In this case, the null hypothesis is that the frequency distribution (or probability model) for the underlying distribution is no different from a normal distribution. The χ^2 test is also referred to as the goodness-of-fit test. We note that the χ^2 test can also be used to compare a sample distribution with other probability models besides the normal distribution.

We have covered one-and two-factor ANOVA. But one will also encounter multivariate ANOVA, in which there is more that one dependent variable (more than one outcome or measure being sampled.) Such analysis is referred to as MANOVA and is used in a more generalized multiple regression analysis in which there may be more than one independent variable and more than one dependent variable. The dependent variables are multiple measures (or repeated measures) drawn from the same experimental units being subjected to one or more factors (the independent variables). In linear regression, we sought to determine how well we could predict the behavior of one variable given the behavior of a second variable. In other words, we had one independent variable and one dependent variable. In the case of multiple regression analysis, we may assume more than one dependent variable and more than one independent variable to explain the variance in the dependent variable(s). For example, we may have one dependent variable, such as body weight, which we model as a linear function of three independent variables: calorie intake, exercise, and age. In performing a multiple regression analysis, we are trying to predict how much of the variability in body weight is because of each of the three independent variables. Oftentimes, one independent variable is not enough to predict the outcome of the dependent variable. However, we need to keep in mind that one should only add additional predictors (independent variables) that contribute to the dependent variable in a manner that the first predictor does not. In other words, the two or more predictors (independent variables) together must predict the outcome, or dependent variable, better than either independent variable can predict alone. Note that a full, generalized linear model allows for multiple dependent and independent variables. In essence, the multiple regression model is looking for simple correlations between several input and output variables.

Although we have focused on linear regression, the reader should also be aware that there are models for nonlinear regression that may be quite powerful in describing biological phenomena. Also, there is a whole literature on the analysis of errors that result when a model, used to predict the data, is compared with the actual measured data [3]. One can actually look at the residual errors between the modeled or predicted data and the actually measured data. Significant pattern or trends in the magnitude and ordering of residuals typically indicate a poor fit of the model to the data. Such patterns suggest that not all of the predictable variability in the measured data has been accounted for by the model.

Other multivariate analyses include cluster, discriminant, and factor analyses. These topics are covered in numerous statistical texts. These analyses allow one to group a population into subpopulations and explain complex data of seemingly many dimensions of factors into a smaller set of significant factors.

In this text, we have also not covered receiver operator characteristic curves. These are statistical analyses used frequently in the design and assessment of the medical devices or diagnostic tests used to detect disease or abnormalities. These curves, often summarizing terms such as *sen-*

sitivity, *specificity*, and *accuracy*, provide a means for determining how accurately a diagnostic tool, test, or algorithm is in detecting the disease or abnormal physiological function. We have previously discussed types I and II errors, which can also be used to estimate the sensitivity, specificity, and accuracy of a diagnostic test. There is often a trade-off between sensitivity and specificity, which can raise frustration for the biomedical engineer trying to develop safe, accurate, practical, and inexpensive diagnostic tests. The receiver operator characteristic curve is a graph that plots the sensitivity of the test (probability of a true positive result) against the probability of a false-positive test. The operator usually chooses to operate at the point on the receiver operator characteristic curve where sensitivity and specificity are both maximized. In some cases, reducing specificity at the expense of sensitivity may be preferred. For example, when administering a test to detect streptococcal throat infection, we may prefer to maximize sensitivity in order not to miss a diagnosis of strep throat. The trade-off is that we may reduce specificity, and a person without streptococcal infection may be misdiagnosed as having the streptococcus bacteria. The result is that the person ends up paying for and consuming antibiotics that are serving no purpose. In other clinical situations, the trade-off between sensitivity and specificity may be much more complex. For example, we often wish to use a noninvasive imaging tool to detect cancerous breast lesions. High sensitivity is desired so that the test does not miss an occurrence of a cancerous lesion. However, the specificity must also be high so that a patient will not have unnecessary surgery to remove healthy or noncancerous tissue.

Finally, there is a vast area of statistical analysis related to time series, that is, data collected over time. Time is the independent variable being used to predict the dependent variable, which in biomedical applications, is often a biological measure. Statistical time series analysis is nicely introduced in Bendat and Piersol [7] and plays an important role in most biomedical research. Such analysis is often used to develop automated detection algorithms for patient monitoring systems, implantable device, and medical imaging systems.

This text alludes to, but does not explicitly cover, the use of statistical software packages for the analysis of data. There are a number of statistical software programs that are commercially available for the engineer who needs to perform statistical analysis. Some of the software packages are freely available, whereas others can be quite expensive. Many of the software packages offer some tutorial assistance in the use and interpretation of statistical and graphical analysis. Some have user-friendly interfaces that allow the user to quickly load the data and analyze the data without much instruction. Other packages require considerable training and practice. Some of the more popular software packages for statistical analysis used in the biomedical field include SPSS, Minitab, Excel, StatView, and Matlab.

Finally, an important takeaway message for the reader is that statistics is neither cookbook nor cut-and-dry. Even the most experienced biostatisticians may debate over the best analyses to use for biomedical data that are collected under complex clinical conditions. Useful statistical analysis requires that the user first frame a testable hypothesis, that the data be collected from a representative sample, and that the experimental design controls for confounding factors as much as is practical. Once the data are collected, the user needs to be as fair as possible in summarizing and analyzing the data. This requires that the user have a good appreciation for the assumptions made about the underlying populations when using a statistical test as well as the limitations of the statistical test in drawing specific conclusions. When used properly, statistics certainly help us to make good decisions and useful predictions, even in the context of uncertainty and random factors over which we have no control. Biomedical engineers need to embrace statistics and learn to be as comfortable with the application of statistics as they are with the application of algebra, calculus, and differential equations.

• • • •

Bibliography

[1]   Task Force of the European Society of Cardiology and the North American Society of Pacing and Electrophysiology, "Heart rate variability. Standards of measurement, physiologic interpretation and clinical use," *Circulation*, vol. 93, pp. 1043–1065, 1996.

[2]   Wagner, G.S., *Marriott's Practical Electrocardiology*, 10th ed., Lippincott Williams & Wilkins, Philadelphia, PA, 2001.

[3]   Hogg, R.V., and Ledolter, J., *Engineering Statistics*, Macmillan Publishing, New York, 1987.

[4]   Gonick, L., and Smith, W., *The Cartoon Guide to Statistics*, HarperPerennial, New York, 1993.

[5]   Salkind, N.J., *Statistics for People Who (Think They) Hate Statistics*, 2nd ed., Sage Publications, Thousand Oaks, CA, 2004.

[6]   Runyon, R.P., *Fundamentals of Statistics in the Biological, Medical and Health Sciences*, PWS Publishers, Boston, MA, 1985.

[7]   Bendat, J.S., and Piersol, A.G., *Random Data. Analysis and Measurement Procedures*, 2nd ed., Wiley-Interscience, John Wiley & Sons, New York, 1986.

[8]   Ropella, K.M., Sahakian, A.V., Baerman, J.M., and Swiryn, S., "Effects of procainamide on intra-atrial electrograms during atrial fibrillation: Implications for detection algorithms," *Circulation*, vol. 77, pp. 1047–1054, 1988.

[9]   Fung, C.A., *Descriptive Statistics. Module 1 of a Three-Module Series on Basic Statistics for Abbott Labs*, Abbott Labs, Abbott Park, IL, December 11, 1995.

[10] Olkin, I., Gleser, L.J., and Derman, C., *Probability Models and Applications*, Macmillan Publishing, New York, 1980.

[11] Fung, C.A., *Comparing Two or More Populations. Module 2 of a Three-Module Series on Basic Statistics for Abbott Labs*, Abbott Labs, Abbott Park, IL, December 12, 1995.

[12] *Minitab StatGuide. Minitab Statistical Software*, release 13.32, Minitab, State College, PA, 2000.

[13] Ropella, K.M. Baerman, J.M., Sahakian, A.V., and Swiryn, S., "Differentiation of ventricular tachyarrhythmias," *Circulation*, vol. 82, pp. 2035–2043, 1990.

[14] Schmit, B.D., Benz, E., and Rymer, W.Z., "Afferent mechanisms of flexor reflexes in spinal cord injury triggered by imposed ankle movements," *Experimental Brain Research*, vol. 145, pp. 40–49, 2002.

Author Biography

Kristina M. Ropella is currently a professor of biomedical engineering at Marquette University, where she has been a member of the faculty since 1990. She received her Doctor of Philosophy degree in biomedical engineering from Northwestern University (Evanston, IL). Her research interests are in physiological signal processing, electrophysiology of heart and brain, functional magnetic resonance imaging, and biocomputing. She teaches undergraduate and graduate courses in biomedical signal processing, statistical time series analysis, biostatistics, biomedical computing, and biomedical instrumentation design. She is a past recipient of the Marquette University Robert and Mary Gettel Faculty Award for Teaching Excellence and the College of Engineering Outstanding Teacher award. She also directs the Functional Imaging Program, a joint doctoral degree program offered by Marquette University and the Medical College of Wisconsin. She is a fellow of the American Institute of Medical and Biological Engineering, a senior member of the IEEE Engineering in Medicine and Biology Society, and a member of the Biomedical Engineering Society and the American Society for Engineering Education. She serves on the editorial board for IEEE Book Press, biomedical engineering series, and she has chaired a number of education- and university industry-related sessions at the EMBS, BMES, and ASEE society meetings. Her personal interests include raising three children, racquetball, dance, travel, and lay ministry for the church.

Printed in the United States
by Baker & Taylor Publisher Services